PATRONAGE AND EXPLOITATION

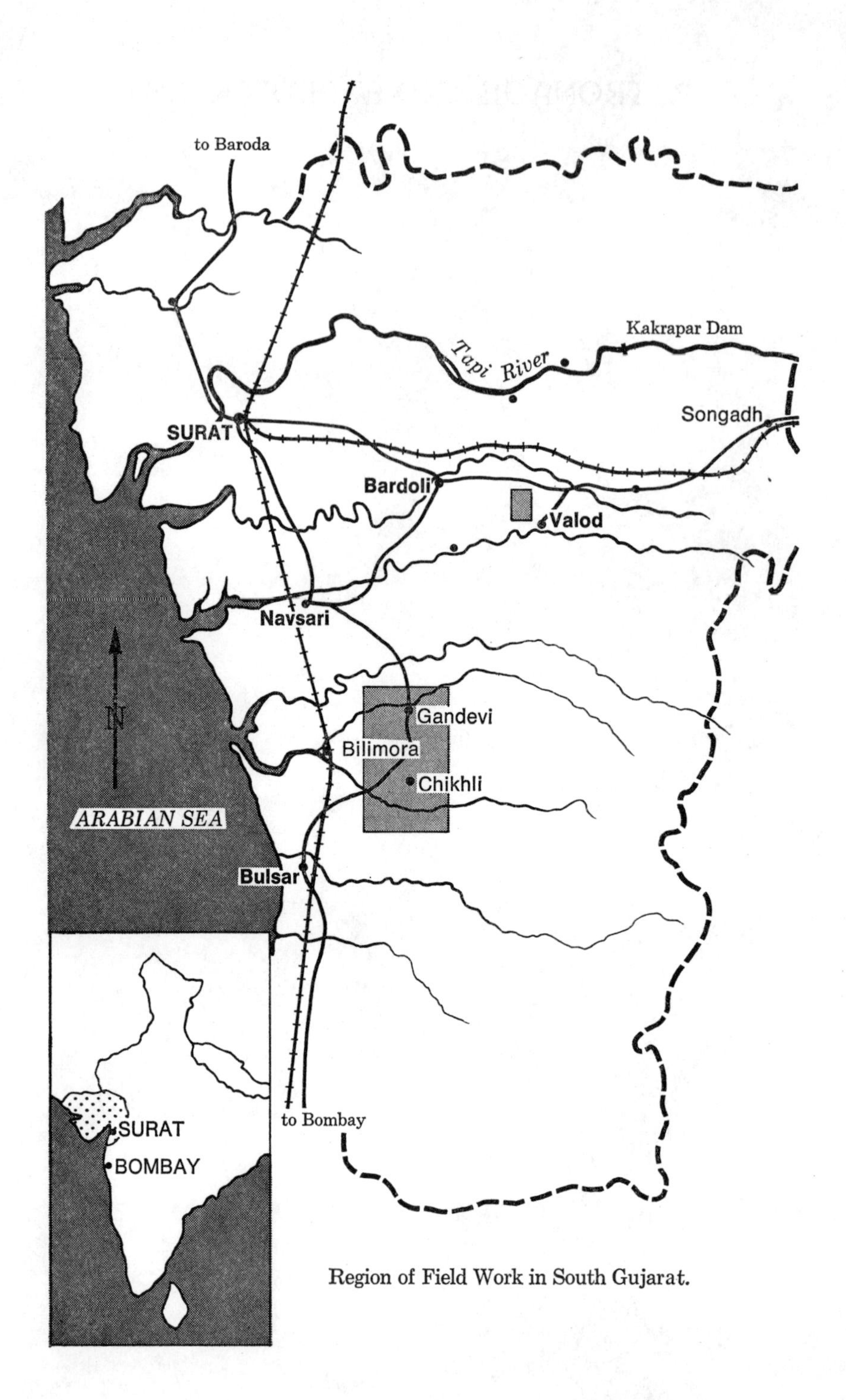

Region of Field Work in South Gujarat.

PATRONAGE AND EXPLOITATION

Changing Agrarian Relations in South Gujarat, India

Jan Breman

UNIVERSITY OF CALIFORNIA PRESS
BERKELEY, LOS ANGELES, LONDON

University of California Press
Berkeley and Los Angeles, California
University of California Press, Ltd.
London, England

ISBN: 0-520-02197-5
Library of Congress Catalog Card Number: 73-186114
Printed in the United States of America

Foreword

Jan Breman's present study originated in a research project carried out by a group of young Dutch sociologists and social anthropologists in southern Gujarat. Among the subjects were the effect of the abolition of tenancy (C. Baks), the impact of governmental planning activities at the local level (E. W. Hommes), and the phenomenon of hypergamy among the Anavil Brahmans, the dominant caste in a large part of the area (K. W. van der Veen). Although the researchers had their individual responsibility and did not operate as a team, yet their studies supplement each other in such a way as to present a more or less coherent picture of south Gujarat rural society.

All of them are highly indebted to the kind cooperation on the part of the Indian people and authorities, received throughout the period during which the researchers were in the field. Professors I. P. Desai and M. B. Desai, at that time both of them attached to the Maharaja Sayajirao University at Baroda, have earned our special gratitude through their never relenting interest in our project and their expert advice liberally given on many occasions. The studies resulting from the project were all accepted as doctoral dissertations at the University of Amsterdam, and published in Dutch in a limited number by the Department of South and Southeast Asia of the Anthropological-Sociological Center of that university. Efforts are being undertaken to make the other studies also available in English translation.

When, in January 1963, I visited the group at work in the area at that time, I once called upon one of the authorities, together with Dr. Breman. Asked about his project, the latter replied that he was concerned with the *hali* system. The official retorted that it was no use investigating that issue, since there were no longer any *halis*— the *hali* system had been abolished long ago. Not taken aback at all, Breman pursued his research project and found many data which are not generally known among Indian officials, and maybe not even among most of the Indian sociologists. I feel that Dr. Breman's study may substantially contribute to our insight into the dynamics of Gujarati society, and perhaps even of Indian society in general.

The "technique" applied by Dr. Breman looks very simple. He did not start with devising a questionnaire and administering it from a nearby city, but settled for several months among the villagers whose labour relationships he wanted to investigate. He followed, therefore, the observation "technique" usual among anthropologists. But in a highly stratified society like the Indian one, investigating a group like the Dubla land labourers presents an additional obstacle. Automatically, a foreign researcher finds his "natural" environment first and foremost among the members of the dominant caste, the patrons in the labor relationships. It is difficult for a researcher not to let himself be lured into looking at the relationships and at the opposite group mostly through their eyes. On the other hand, a researcher may be identified, by the servant group, with the people among whom he was mostly seen during the first weeks, and with whom he has several traits in common.

It seems to me that the way in which Breman approached the Dublas, and his serious attempts to complete his picture of Gujarati society by adding their perspective to the one representing the patrons' view, form the core of his "research technique." It is this aspect of his research which lends this research the character of a truly modern study.

I am convinced that as a pioneering approach to a society that has never been thoroughly studied, participant observation is a much more promising "technique" than administering questionnaires in order to collect quantitative data, which in such a situation mostly amount to a false pretense of exactitude.

Another aspect of the present study which deserves special mention is Dr. Breman's attempt to place his findings within a theoretical frame. Although it is not the first time that Indian land labor and *jajmani* relationships have been analysed in terms of patronage and clientele, it seems to me that Breman's sociohistorical analysis has added a new dimension to the way social change within Indian rural society should be interpreted in order to obtain a diachronic perspective.

A final point I would like to mention is the relevance of the crop pattern and of the general attitude of dominant caste members to employment alternatives among their land laborers. It is a highly interesting finding that despite the nearness of a small town the Dublas in one of the two villages investigated were not able to find the way towards urban employment, simply because the Anavils did not want them to leave their village, since they needed cheap labor for their mango gardens.

Finally, I would like to express my thanks to the Netherlands Organization for the Advancement of Pure Research for the liberal way in which, from the very first beginning, they sponsored and financed the south Gujarat research project. I also want to thank my colleagues Dr. A.J.F. Köbben and Dr. S.C.L. Vreede-de Stuers for their assistance in setting up and guiding the whole project.

W. F. Wertheim
Department of South and Southeast Asia
Anthropological-Sociological Center
University of Amsterdam

Contents

TABLES

Introduction

India has retained its character of a peasant society despite the accelerated process of urbanization and industrialization which it has undergone in the last few decades. As the census of 1961 showed, 82 percent of the population lives in rural areas, and roughly 70 percent is engaged in agriculture. These figures in themselves more than justify the continued interest of sociologists and anthropologists in India's rural structure.

Nevertheless, until recently only a limited part of the abundant literature on village India was concerned with agrarian sociology. A major preoccupation has been to study the caste society as a clear-cut system of social organization, while the network of social relationships involved in the organization of agrarian production has long remained of secondary interest.

Although it was generally recognized that India is not one of those peasant societies with little differentiation as to ownership and utilization of land on the village level, information on the division of the local community into landowners, small farmers, and agricultural laborers often had to give way before detailed description of the variations among the artisan and serving castes. What surprises us is that, in the sometimes very detailed listings of the activities of the various castes, agricultural labor — surely the principal activity in a peasant economy — is scarcely mentioned, except as work which is not very specific and can be performed by members of all castes.

Moreover, in many publications based on field work the propertied high castes in a village received proportionately more attention than the others. Concerning the members of the lower castes, often landless, data are much more scanty. As a result, we know comparatively little about the conditions in which the poorest people in India work and earn a living. They are the agricultural laborers, who are estimated to constitute approximately 25 percent of the agrarian population. It is true that two successive government reports — the records of the *Agricultural Labour Enquiry* of 1950–51 and 1956–57 — have made available a great deal of information in regard to this category, but the material they contain has not been interpreted in a wider framework.

My study is concerned with landowners and agricultural laborers in the southern part of the State of Gujarat. It is based on a combination of source studies and field-work reports. The relations between Anavil Brahmans, the principal local landowners, and Dublas, a tribal caste of landless laborers, still contain many elements of the form of service which existed in earlier times and which is generally known as the *hali* system.

For an understanding of the present situation an insight into the position of the agricultural laborer in the past is indeed of essential importance. After my field-work period, this conclusion led me to a study of the historical sources, which would enable me to analyze the master-servant relationship in the traditional society. Thus, the results of my investigation into the existing situation formed for me the starting-point for an interpretation of the past relationships between the two groups, and not the other way round. This made my book difficult to plan, but in the end I decided to arrange the chapters in chronological order, so that the material leads from the general to the particular. By employing data that go back to the beginning of the nineteenth century I have tried to record and explain the successive changes in the relationships between landowners and landless laborers, two important categories in the rural structure.

In order to place my argument against a more general background, I have begun this study with a brief outline of the development of agrarian labor relationships in India. This introduces a theoretical discussion of the relationships between landowners and agricultural laborers in the light of the *jajmani* system, the framework in which the intercaste relationships in the traditional society were institutionalized (part I). A brief description of the country and people of south Gujarat is then followed by a summary of the *hali* system as it functioned during the last century in the region under discussion, and of the changes which have occurred since then (part II). The conclusions from this part of my account are at the same time a starting-point for the report of the investigation into the existing relationships between landowners and agricultural laborers which I carried out in two south Gujarat villages during 1962 and 1963 (part III).

While my aim during field work was to gather data on the present situation, I supplemented my study by examining historical material after my return from India. A subsidy from the Netherlands Organization for the Advancement of Pure Research in 1966 enabled me to consult published and unpublished documents in a number of London libraries, and the same organization made possible the present English translation by Mrs. Wil van Gulik from the original Dutch version published in 1970.

The continuous exchange of ideas with Professor Wertheim—before my departure, during his visit in the field, and after my return—has clarified for me the concept of *guru.*

The friendly cooperation in India with Chris Baks, Enno Hommes, Igle Ronner, and Klaas van der Veen — my fellow-participants in the Gujarat project—has been of great importance to me, and a source of inspiration.

Miss H. L. Lambert, M.A., who was at the time attached to the London School of Oriental and African Studies and taught me Gujarati, made it easier for me to conduct my investigation and enabled me to be a little less of an outsider in the community with which I associated for a time.

Without wishing to understate the help of all those who assisted me in India when I was carrying out the investigation, I feel that special thanks are due to Professor I. P. Desai and Professor M. B. Desai. Their professional qualifications (sociology and agricultural economy, respectively) and their personal familiarity with the region under research, resulted in advice and suggestions that were very valuable to me. I am, moreover, indebted to Professor and Mrs. M. B. Desai for the hospitality and assistance they gave my wife during her stay in Baroda.

Professor K. de Vreese of the University of Amsterdam was so kind as to check the Glossary for imperfections. I would also like to express my appreciation to Professor Köbben for his careful reading of the manuscript. The final result bears to some extent the stamp of his many critical notes.

Finally, I gratefully acknowledge the aid I received from my wife. She corrected the final text, but, much more important, in the earlier stages of gathering and writing up materials she put me right many times.

"It is my duty . . ." In this way the Anavil Brahmans of south Gujarat deprecate expressions of gratitude, even if they wish to leave no doubt of the obligations under which their interlocutor has just come. It gives me pleasure that I am able at last to honor these obligations. Dutiful acknowledgement is, however, the last of the reasons why I have mentioned a number of people by name who helped me in the course of my study. Nevertheless I take this opportunity to express my most profound thanks to all those who enabled me in the field to collect the data of which this book is the result.

After an interval of seven years, I returned to south Gujarat in the summer of 1971 to carry out a new research project: an analysis of the institutional framework of unskilled labor and of the social mobilization of rural and urban laborers in the district. I used this opportunity to revisit the scenes of my earlier field work and to renew my association with the people I had known before. I was especially interested in finding out how the "green revolution" had manifested itself in the area.

In the Postscript, written after my return to the Netherlands in 1972, I comment on continuity and change in the relations between landlords and agricultural laborers in south Gujarat.

Jan Breman
Faculty of the Social Sciences
Erasmus University
Rotterdam

PART I

Chapter I

Agricultural Laborers in India

1.1 The Development of Agrarian Labor Relationships

Recent data show that nearly a quarter of the agrarian population of India consists of agricultural laborers, a category long supposed to have been practically nonexistent at the beginning of the nineteenth century. In the literature of economic history, the precolonial society was usually depicted as a politically autonomous, economically self-sufficient, and more or less static community. Peasant castes on the one hand, and artisan and serving castes on the other, collaborated on a footing of equivalence and reciprocity. The peasant castes tilled the soil and exchanged part of their crops for the goods and services of the nonagrarian specialists. In this conception of the homogeneous village community, agricultural laborers did not figure, or at most in negligible numbers.[1]

The rise of a predominantly landless category of laborers was attributed until recently to changes in the economic structure of the local community under colonial rule. A complex of factors was thought to have caused a large-scale decline down the agrarian ladder, a process for which colonial rule was held responsible: peasant landholders became tenants and finally agricultural laborers.

British rule and the flooding of India with foreign manufactures destroyed domestic industries, and so drove the artisan on to the land. The British introduced in certain areas a system under which land revenue was assessed at high rates and was payable in cash, and which held individuals responsible for payments. This led to the destruction of the old village communities. The British brought about changes in the law which made

1. For many years this was the prevailing view. S. J. Patel summarized it as follows: "In pre-nineteenth century India . . . there were domestic and menial servants; but their numbers were small and they did not form a definite group . . . The large class of agricultural labourers represents a new form of social relationships that emerged during the late nineteenth and early twentieth centuries in India." (Patel, 1952, 32.)

it possible to sell land; the monetization of the economy, the stagnation of agriculture, and the pressure of population on resources increased the propensity to do so. The peasant was forced to sell land; it went either to the State for non-payment of taxes, or to the money-lender for non-payment of debt. This turned the peasant into a landless labourer.[2]

In these terms, D. K. Kumar has summarized the ideas that were long current in regard to the changes in the social and economic order since the beginning of colonial rule. On the basis of her research into the sources on the situation in south India in the early nineteenth century, she was the first to challenge that view in a fundamental way.[3]

Of course, the political, economic, and demographic changes of the nineteenth century did influence the relations of power and property in the rural areas. It is also quite possible that the percentage of agricultural laborers increased during that period, although the obscure and varying classifications of this category of labor in the successive censuses make it difficult to reconstruct the progress of this increase.[4]

During the colonial era there was undoubtedly a process of growing indebtedness and, in densely populated areas, increasing subdivision of the land. Did this, however, really lead to an agrarian regression which finally resulted in a proportion of agricultural laborers that has been rising ever since the previous century? It is impossible to analyze changes in the agrarian structure as a purely economic process. In peasant societies, changing relationships in landholding are reflected in changing positions in the social structure. If a substantial part of the rural population had drifted into a life of landless agricultural labor, a large-scale process of social descent would have resulted. There are, however, no signs to show that earlier generations of those who are now agricultural laborers, usually belonging to the lowest castes, occupied higher positions in the caste hierarchy at the beginning of colonial rule. Various non-agrarian castes worked on the land as well, and often it was their chief occupation, while they performed their caste functions — tanning, sweeping, weaving — incidentally. Moreover, in many areas

2. Kumar, 1965, 188.

3. In many earlier publications the existence of categories of agricultural laborers in precolonial India is indeed pointed out. See, for instance, Moreland, 1920, 111–5; R. Mukerjee, n.d., 62; Kosambi, 1956, 353; Neale, 1962, 22f.

4. See, for instance, Thorner, 1962, ch. 10; Kumar, *op. cit.*, ch. 10.

members of the low castes were excluded from landownership in order that their working power could be mobilized.[5]

Kumar calculated that, already in the early nineteenth century, agricultural laborers in south India formed at least 17 to 25 percent of the agrarian population,[6] and the situation in that area was no exception. An analysis of the first settlement reports drawn up by British district officials in the last century leads to the same conclusion for the area of south Gujarat.

That nevertheless so little attention was paid to the landless categories among the agrarian population is due to various factors. In the early colonial historiography, discussion of the Indian agrarian structure was largely restricted to the questions of who the landowners were and how they should be taxed. Nor does the more recent literature of economic history mention much differentiation as to landownership among the various castes in the precolonial period.

No doubt the picture of a closed, more or less harmoniously functioning "community" of peasants and nonagrarian specialists has been greatly influenced by the anticolonial and anticapitalist viewpoints that were brought to the fore in the struggle for national independence.[7] The image of the idyllic village community, characterized by reciprocity instead of competition, by solidarity instead of exploitation, compared favorably with the structure of the rural community as it was said to have come into being under British rule. This stereotype of a better past is still to be found in the ideology of the community-development organization, in which a plea is made for a return to the original village democracy based on harmonious cooperation and egalitarian relationships.[8] Although indeed in recent years this version has receded into the background, the ideological bias has not, a case in point being the decision to establish *panchayat* councils as the lowest politico-social units.

There is sufficient reason to suppose that, even at the beginning of the nineteenth century, a substantial part of the population con-

5. Pandya, 1959, 155; Kumar, *op. cit.*, 31.

6. Kumar, *ibid.*, 191.

7. Dumont, 1966, 67–89, and Neale, *op. cit.*, 45–47, were among those who conclusively refuted this representation of the precolonial village community, characterized in a more general sense by Wittfogel as a "beggars' democracy" (Wittfogel, 1963, 123). Such distortion of the past occurs both in the nationalist historiography on the traditional agrarian structure and in the Indian Marxist literature on the subject. For the latter, see, for instance, A. R. Desai, 1961, and B. Sen, 1962.

8. The writings of the former Minister of Community Development S. K. Dey are an example.

sisted of agricultural laborers. The differentiation among the agrarian population long antedates the colonial era. Investigators have come to realize that rights to the land were unevenly distributed among the various castes at a very early time. Agricultural laborers and members of the nonagrarian castes, who by way of compensation for services rendered could temporarily cultivate plots of land, should not be termed landowners. The complex character of the agrarian labor relationships and their wide regional variety make any categorization difficult. But to state, as is often done, that in the past agriculture as a sole occupation was open to all castes is to say that the members of the lower castes depended on the landowning castes for their living. They carried out all kinds of defiling and otherwise less-favored work and depended, whether as tenants, share croppers or agricultural laborers, on the higher castes, in which the bulk of ownership had always been concentrated.[9]

In this study, the total pattern of dependence resulting from an unequal distribution of landownership has not been considered, for I am concerned exclusively with the relations between landowners and agricultural laborers as they have developed since the last century.

1.2 Agricultural Laborers in the Traditional Society

So far as any information on this subject can be found in the literature, it stresses the personal servitude of the members of the agricultural-laborer castes in the past. Many of them were permanently bound by debt to members of the landowning higher castes, by whom they were employed indefinitely.

Bondage existed in many gradations and was not restricted to agricultural laborers. Small farmers who entered into an agreement of tenancy or crop-sharing undertook to hand over part of the crop, but also to perform field services (*begar*), hence they could not till their own plots at the most suitable moment. In various ways and in varying degrees the landowning castes managed to establish complete control over the socially and economically weaker groups of the village and thus to monopolize their labor permanently or periodically. In general it can be said that the condition of dependence, on which the agrarian relationships were usually founded because

9. Leach, 1967, 12.

of the concentration of landownership, led to more or less permanent or even hereditary attachment. Nevertheless, bondage and permanence were preeminently characteristic of the category of farm servants, for whom they bore a far more personal character. It was in this category that the dependence of the landless laborers on the landowners was most pronounced.

Were there other categories of agricultural laborers aside from these bonded servants? Moreland thought not.[10] Kumar, in the beginning of her study, remarks incidentally that, in south India in the early nineteenth century, labor relationships varied from slavery to completely free labor. But the material she presents relates predominantly to relationships of bondage. Other publications also show that in many regions of India hereditary servitude, institutionalized in bondage, was the condition of a very large proportion of the caste of agricultural laborers.

The bondsmen were known as *halis* in Gujarat; *padiyals, pannaiyals* and *charmas* in Madras; *adimas* in Kerala; *huttalus, mannalus,* and *jeethas* in Mysore; *busaliyas* and *shalkaris* in Madhya Pradesh; *bhagelas* and *gassi-galus* in Andra Pradesh; *muliyas, gothis, chakars,* and *haliyas* in Orissa; *kuthias, krishans,* and *chakars* in West Bengal; *kamias, baramasiyas,* and *janouris* in Bihar; *harwahas, hariyas,* and *sewaks* in Uttar Pradesh; and *halis* and *sepis* in Punjab.[11] They were often referred to by their caste names, an indication that bondage was indeed the predominant condition of agricultural laborers. They belonged without exception to the untouchable or tribal castes, the lowest in the hierarchy.

> Agrestic serfdom is most prevalent in those parts of India where the lower and depressed orders are most numerous. Bombay, Madras, Malabar, Cochin, the Central Provinces, Berar, Central India and Chota Nagpur show the largest aboriginal population, and it is in these areas that the status of agricultural labourers verges most nearly on slavery.[12]

As for a large part of the tribal castes, it is doubtful whether in the Hindu social constellation they ever belonged to the class of landowning peasants. Their entry at a low level into the caste system inevitably led to their being taken up in the agrarian production process in a relationship of dependence, that is, as agricultural la-

10. Moreland, *op. cit.*, 115.
11. D. Desai, 1942; Sivaswamy, 1948; Lorenzo, n.d.; Singh, 1947; and Beidelman, 1959, 10.
12. R. Mukerjee, 1933, 225f.

borers or tenants. If their ancestors had had claims to the areas in which they wandered as "shifting cultivators," these claims were lost when their descendants were integrated into the caste system. This process of incorporation went on slowly, taking place in various parts of India at different times. In some regions it ended long ago, whereas in others it did not begin before the colonial era.

As agricultural laborers they were everywhere employed by members of high castes, who owned a disproportionately large part of the land. Landlords is a suitable term for these propertied peasants, who often owed it to their influential position and high ritual status to perform little or no agricultural labor themselves.[13]

Landlords and agricultural laborers, in other words, represented the extremes of the economic structure and of social stratification in the rural areas. A greater difference than that between the seignorial style of life of the richest and most powerful people in the village and the situation of poverty and bondage of the farm servants is hardly conceivable.

The servitude of the agricultural laborers was institutionalized in different ways from region to region, but the multiplicity of forms cannot conceal that the type of relationship was essentially the same everywhere. The dependent relationship usually came into being when a member of a caste of agricultural laborers accepted bondage in exchange for a loan in kind or in cash, most often to meet the expenses of his marriage. He undertook to work as a farm servant for a landowner until he had paid off his debt. But because of the low compensation he received for his services, his chances of discharging his debt were extremely slight. The debt tended to increase in the course of time, and the farm servant, with few exceptions, thus remained in bondage for the rest of his life. The attachment was continued by the next generation when the son of a farm servant also married at the expense of his father's master. In this sense the agreement between landlord and agricultural laborer was hereditary as well as permanent.

The origin of bondage has been explained in different ways. Adhering to the idea of a process of agrarian proletarization under colonial rule, S. J. Patel suggested that the members of the artisan and serving castes were driven by economic necessity to accept bondage as the traditional village community disintegrated. He ar-

13. Kumar, *op. cit.*, 189f; Lorenzo, *op. cit.*, 75; Singh, *op. cit.*, 23f; and Ramakrishnan, 1948, 18.

gued that this individual bondage took the place of the group bondage of the precolonial period.[14]

His assumption that, in the past, members of some castes were collectively bound in service to groups of landowners is nowhere confirmed by fact. Bondage has a long history; it existed on a wide scale in the Moghul period. When a region was opened up, the tribal population presumably lost their control of the land and were compelled to accept bondage to the incurrent caste Hindus. In some regions this process was still going on in the recent past, hence this seems a feasible reconstruction.

An analysis of the factors determining the relationships between landowners and agricultural laborers offers more scope for insight than do considerations of a more or less speculative nature on their causes. In other words, it is more to the point to know the societal background against which the servitude of the agricultural laborers functions than to reflect on the question of when and how it arose.

14. S. J. Patel, *op. cit.*, 87f.

Chapter 2

Landlords and Agricultural Laborers in the Past: Parties in a System of Bondage and Patronage

2.1 Agrestic Serfdom

Servitude as it existed in the past has been generally regarded as a form of unfree labor. The farm servants are invariably referred to as "landless serfs," "bonded servants," "agrestic slaves," "debt servants," and the like. The relationship of coercion derived mainly from the debt that originally led to the attachment, from the servant's inability to sever the relationship, and from the more or less automatic continuance of the bondage from father to son.

It has often been argued that servitude is associated with a need for regular laborers. In this view, it was the largest landowners, doing little or no work on the land themselves, who imposed servitude. Only in this way could they be sure of a steady supply of labor. When new regions were opened up, the local tribal population was subordinated by the incurrent Hindu groups.

> Where agriculture is practised with crude implements and without the aid of domestic animals, where the working population is scarce, where the land must be reclaimed from the wilds and marshes, and where soil and climate act as limiting factors for the employment of imported labour — it is not capital that is wanted, but native labour to reclaim the land and cultivate it under difficult environmental conditions. Under these circumstances bond-labour of the native population is introduced and it is pinned to the soil in conditions akin to slavery and serfdom.[1]

The theory formulated by Nieboer on the occurrence of slavery, extended by Kloosterboer to other forms of unfree labor, is completely

1. Lorenzo, 1945, 132.

in keeping with this interpretation.[2] For where it is possible for the cultivator to be self-supporting—that is to say, where land is not in short supply—no voluntary labor will be forthcoming, and forced labor in one form or another is inevitable.

Srinivas adduces arguments in support of that theory. He observes that in precolonial India there was a scarcity of agricultural laborers, and that dependent low castes could take refuge in as yet unreclaimed regions.[3] In such circumstances, forced labor flourishes. Conversely, the disappearance of servitude in its original form during this century is consistent with the theory. The increase of the population of agricultural laborers, it is suggested, decreased the need for landowners to apply labor compulsion.[4]

This is a decidedly one-sided presentation, and in direct contrast to the view that bondage was not associated with a lack of voluntary labor, but with a lack of employment, to which agricultural laborers reacted by accepting voluntary servitude.[5] In that case, the initiative was taken by the economically weak, who tried to find security by putting themselves in the hands of a landlord. For them, too, the sharply fluctuating demand for labor was an incentive toward accepting permanent bondage. A bondsman's son was entitled as well as obliged to apply to his father's master for a loan, and in this way the existing tie was prolonged into the next generation.

> It was not unknown for the descendants of labourers to claim work as a right from the descendants of former employers. After all, there were certain advantages about servitude; from one point of view, an obligation to work could be construed as a right to employment.[6]

To this it should be added that, as the rights in land were monopolized by a colonizing group, the local tribal population was practically forced to comply with terms of service which drastically restricted their freedom.

In this conception, it is true, the bondage of the agricultural la-

2. Nieboer, 1910, 2nd ed.; Kloosterboer, 1954. For a reinterpretation of Nieboer's theory, see Baks, Breman, and Nooy, 1966.

3. Srinivas, 1966, 42f.

4. Dovring, in an article on the economy of unfree labor, writes, "Landlords had an interest in bonded peasantry as long as labor was scarce in some sense they could experience. Liberation came above all when there was no longer scarcity; in the latter situation, economic pressure would replace the previous lack of civil freedom." (Dovring, 1965, 321.)

5. This is the essence of Patel's argument concerning the rise of categories of agricultural laborers under colonial rule (S. J. Patel, 1952, 88).

6. Kumar, 1965, 45.

borers resulted from their inability to earn an independent living, but it rests on the assumption that the dependent category was unable to withdraw from the authority of the dominant landowners by geographical mobility. The openness of the agrarian system, which plays a decisive part in Nieboer's theory, is in this view more apparent than real. In the regions that had already been reclaimed, at any rate, bondage was not simply imposed. While as a rule it is the way in which the relationship has come into being that is the criterium of unfree labor, for the attached servant the unfree element rested rather in his inability to break the bond once it had been contracted. For this reason traditional bondage was a special form of unfree labor, best described as voluntary servitude.

The last interpretation is one that does not fit the assumption of a lack of labor supply in the past. Only when employment outside agriculture became possible was the agricultural laborer's need of security lessened and, with it, his willingness to accept permanent bondage. This explanation throws a different light on both the incidence and the disappearance of bondage among the agricultural labor population.

In representing bondage as a form of unfree labor, investigators have heretofore paid attention only to factors arising either from a need of laborers, leading to forced attachment, or from a need of security, leading to voluntary servitude. These factors have in common that they explain bondage as it existed in the past on exclusively economic grounds and purely as a relation of labor.[7]

In my discussion of the *hali* system in south Gujarat I shall demonstrate in more detail that neither interpretation covers all the facts. The relationship between landlords and agricultural laborers as it developed in the course of time cannot be dissociated from the total context of the traditional Indian village or from the changes that have occurred within it. That is to say, the relations between landowners and farm servants must be judged in the light of a much more general pattern of relationships, which prevailed among the various castes in the traditional society and which has come to be called the *jajmani* system.

7. "Part of the reasons why bondage existed or did not exist are outside of the economic condition of the given moment. . . . [It is an] arrangement needed to maintain the sacral, hierarchical, or political institutions which, as we know, often are ends in themselves and not altogether subordinated to considerations of economic welfare." (Dovring, *op. cit.*)

2.2 *Agricultural Labor and the* Jajmani *System*

In studies of the *jajmani* system, emphasis is usually placed on the caste structure as a system of social stratification and of division of labor founded on the distinction between purity and impurity. In the past, every caste performed specific tasks, and in principle it enjoyed the monopoly in the region of carrying out the economic function allotted to it. The resulting interdependence among the castes was expressed in a local network of exchange relationships, which W. H. Wiser termed the *jajmani* system.[8]

The usual definition of this system as a mechanism of exchange of goods and services among the various castes on the local level would, at first sight, comprise the relation between landlords and agricultural laborers as well. Yet many authors do not regard the latter as partners in the *jajmani* system,[9] because, for one thing, the differentiation within the agrarian castes was not sufficiently recognized. This point of view was based on the conception of the traditional Indian village as a community of peasants tilling the soil along with a large number of specialist castes that were nonagrarian. Under these conditions, the exchange of goods and services must necessarily have been restricted to landowners on the one hand and artisan and serving castes on the other.[10]

There was, however, a more fundamental argument to the effect that the *jajmani* system did not include agricultural laborers. It was formulated concisely by D. F. Pocock:

> I would suggest that the heterogeneity of service justifies us in distinguishing one category of specialists to which, by virtue of its religious associations, the term *jajmani* may be applied. These are the specialists whose specialization derives from the exigencies of the caste system and not from economic needs or from the intricacy of a craft.[11]

He reasoned that the relationship between landlords and agricultural laborers was purely economic, on the assumption that this kind of relationship was outside the ideology of the caste system of which the *jajmani* network formed the structural sediment. Even though the members of some of the low castes undoubtedly worked pre-

8. Wiser, 1936.
9. For instance, Cohn, 1955, 56; Gough, in her review of Beidelman's study (see below), 1960; Gould, 1964; and Harper, 1959.
10. The *jajmani* system is explained in these words in Wolf's excellent exposition of the social and economic structure of the peasant society (Wolf, 1966, 38f.).
11. Pocock, 1962, 85.

dominantly as agricultural laborers, their function for other castes has been interpreted primarily in the light of their more specific caste occupations, such as leatherworking or waterbearing, that were exclusively reserved for them.

Although Pocock was right in arguing that the *jajmani* system was not simply an economic structure based on the principle of division of labor, his proposal to speak of *jajmani* relations only in the case of a few — the so-called religious — specialists is in essence equally restrictive. It excludes not only the agricultural laborers, but also the members of the artisan castes, who surely, as a rule, are regarded as having formed part of the system.[12]

To controvert this argument, it can be said that, in a peasant society with a limited functional autonomy of the various spheres, it is impossible simply to separate the performing of religious services from the network of social relationships. It is difficult to distinguish, as Pocock suggests, between the ritual services performed by some castes in support of the norms and values of the ritual hierarchy, and the work carried out by artisan and agricultural castes.

This is notably the case as regards agricultural labor. Pocock is mistaken in thinking that a *hali* in south Gujarat was taken into service only because the landlord needed an extra laborer.[13] No doubt such a need was an important consideration, but it was one that was also determined by the circumstance that agricultural labor — and especially handling the plow — was thought to be defiling. By leaving this kind of work to someone else, the landowner could enhance his prestige. Moreover, in many regions the agricultural laborer traditionally fulfilled some functions of a ritual nature in the house of the landlord.[14] In these cases there was question of a *jajmani* relation even by Pocock's rigid criteria.

As T. S. Epstein observed, it was only in regions with an outstandingly differentiated caste structure that the *jajmani* system comprised a broad diversity of castes.[15] In many parts of India, agricultural laborers fulfilled on behalf of the landlords widely varying tasks which elsewhere were carried out by members of different castes.

The nature of the services, and therefore the contacts among the castes, varied. There were great differences in the extent of speciali-

12. This is also the view of Vreede-de Stuers, 1965, 314–22.
13. Pocock, *op. cit.*, 89.
14. Srinivas, 1965, reprint, 41; Epstein, 1967, 233; Van der Veen, 1972.
15. Epstein, *op. cit.*, 232.

zation, in the exclusivity of the work, and in the frequency with which relations were maintained, but no clear line divided purely secular activities from ritual services. On the contrary, they were interwoven and embedded in a pattern of relationships, involving a number of castes which varied according to local differentiation.

Instead of a limited interaction among a small number of castes, the *jajmani* system had much wider implications, as Gould and others pointed out.[16] High castes tried to avoid low castes because the latter performed defiling or otherwise less favored work, but for that very reason they could not do without them. This dilemma was solved by the *jajmani* system. As the contacts were formalized the distance between the castes was accentuated and bridged at the same time. In this conception, the *jajmani* system was the institutionalization of the intercaste relationships, the mechanism regulating the intricate network of rights and obligations among members of different castes.

By refraining from doing defiling work and, more generally, from engaging in less favored activities, and on the other hand by calling upon the services of highly esteemed specialists, the members of a caste can both enhance their ritual status and emphasize their power and prosperity. Various dimensions of social position are involved and, though differing in weight for each caste, they are closely related. Agricultural labor, for instance, is regarded in many regions as an inferior activity, but only a small group is enabled by its wealth and predominance to live in accordance with this norm. As will appear in greater detail below, when landowners cease to perform agricultural work this is not only taken to be an indication of increased prosperity but also expresses a claim to higher ritual status and authority and, more generally, an attempt to acquire more esteem. Briefly, the *jajmani* relationships expressed differences in purity, material well-being, and power. It is these elements which, sometimes strengthening each other and sometimes mutually opposed, together determined the social positions of the members of a caste in the hierarchy.

This explanation leads to an interpretation of the *jajmani* system as a network of social exchange among the members of the various castes which had religious and economic as well as political facets.[17] However, knowledge of the exact nature of the *jajmani* system and

16. Gould, 1958, 431.

17. Beidelman, 1959, 30; Epstein, *op. cit.*, 15–16; Ishwaran, 1966, 39, and others.

the parties in it does not imply insight into the ways of exchange. It is this aspect that is essential for an understanding of the situation of the agricultural laborers in the past.

2.3 Landlords and Agricultural Laborers: Partners in the Jajmani *System*

Opinions are divided as to the structure of the exchange relationships and the norms and values on which it was based. Wiser emphasized the complementary character of the occupational tasks, thus creating an impression of great role flexibility. The members of the castes were each in turn *jajmani*, receiving performances in the shape of goods and services, and *kamin*, providing such performances. In short, Wiser described a combination of exchange relationships among more or less equal partners. He claimed that the exchange was based on equality and reciprocity, although his factual description of the relations showed that he was not unaware of their asymmetrical character.[18]

Diametrically opposed to Wiser's view is that of Beidelman, who was convinced that inequality of power, based on landownership, was characteristic of the *jajmani* system and decisive of the way in which the exchange came into being.[19] He describes a relationship pattern with a component of exploitation, which obtained between landowning castes on the one hand and landless castes on the other. In general, the members of the ritually high castes were landowners and those of the ritually low castes were tenants or agricultural laborers. The roles of *jajman* and *kamin* were not flexible, but fixed. In the first instance, it was political and economic power, not ritual status, that determined what role was fulfilled by whom. The members of the locally dominant caste were the *jajmans par excellence*, the main landowners and at the same time the political power elite in the village or the region. They did not necessarily belong to the ritually highest or numerically most important castes.[20]

18. An often cited passage from Wiser's study is, "In turn each of these castes has a form of service to perform for the others. In this manner the various castes of a Hindu village in North India are interrelated in a service capacity. Each serves the others. Each in turn is master. Each in turn is servant." His remarks on the services provided by a number of individual castes, for instance the Chamars, testify to a more discriminating judgment. (Wiser, *op. cit.*, 10, 51–53.)

19. Gould, in a publication of 1958, arrived at a similar conclusion, which he retracted in an article of 1964.

20. For the concept of dominant castes see, for instance, Srinivas, 1955; Mayer, 1958; Béteille, 1965.

The members of the other castes depended as *kamins* on this caste of landowners, a dependence which varied with the economic importance and ritual status of the castes concerned. By virtue of their control of the soil, and thus of agrarian production, the members of the dominant castes could claim a disproportionately large part of the total package of goods and services for themselves. The inequality of power that was inherent in the *jajmani* system implied compulsory exchange on unequal terms.

The representation of servitude as a form of unfree labor fits completely into this interpretation. Beidelman regarded bondage and subservience as characteristic of the relation between *jajman* and *kamin* and wrote, not incorrectly, that there was little to distinguish a *jajmani* relation from servitude.[21] For him, the figure of the *jajman* coincided with that of the landlord, and he tended to identify the *kamins* with tenants or agricultural laborers.

Beidelman's study gave rise to a spirited discussion. Many authors agreed that members of the dominant castes occupied central positions in the network of exchange relationships.[22] But not all the members of the local elite were equally prominent, and in fact differed widely in wealth and power. The poorest and least influential among them could not afford to employ a large number of *kamins*, whereas, conversely, there were members of the nondominant castes who, on the strength of their position, managed to evade their obligations as *kamins* and themselves attempted to achieve *jajman* status with all the attendant privileges and advantages.[23]

It has not generally been recognized that the exchange was founded on exploitation. Several authors pointed out the paternalistic attitude of the *jajmans* towards their *kamins*. The *jajman* was expected to behave as a father does towards his son. It was incumbent on him to aid his *kamins*, and this obligation was religiously sanctioned by such values as liberality and magnanimity. The compensations which the *kamins* received were not seen as payment for the labor they had supplied, but as results of the responsibility which rested on every *jajman* for their welfare. This explains why the proportion of the produce distributed among them by the *jajman* at harvest time remained equal, although the sum total of the goods

21. Beidelman, *op. cit.*, 10f.
22. For instance, Pocock, 1957, 79; Gould, 1958, 431.
23. Beidelman, *op. cit.*, 17f.

and services he had received fluctuated. There was no question of a *quid pro quo*.[24]

The *kamins* had not only the duty but also the explicit hereditary right to serve their *jajman*, which guaranteed their basis of existence in an economy of scarcity. Since a paternalistic orientation, on the other hand, implies filial recognition of the father's authority and loyal behavior towards him, the *kamin* owed his *jajman* obedience, deference, and support.

In this description of protégé and protector, the element of patronage is not far to seek. By patronage I mean a pattern of relationships in which members of hierarchically arranged groups possess mutually recognized, not explicitly stipulated rights and obligations involving mutual aid and preferential treatment. The bond between patron and client is personal, and is contracted and continued by mutual agreement for an indeterminate time.[25]

A typification in this terminology makes sense only if by the *jajmani* system, as it functioned in the traditional society, we understand a mechanism not only for the distribution of goods and the exchange of labor but also for the allocation of power and prestige. It is interesting to note that Max Weber already mentioned the *jajmani* principle as a mechanism of patronage.[26]

In most regions of India, a lack of external outlets made local distribution of agricultural products largely inevitable. Any view of patronage in the traditional Indian village should take this fact into account. Maximalization of income remained subject to the increase of esteem and influence of the patron, and this he could not achieve without the support of his clients. As members of the higher castes, the dominant *jajmans* used the services of *kamins* to maintain or enhance their prestige, for instance by dissociating themselves explicitly from certain activities or, conversely, by laying claim to the help of highly favored specialists. As P. M. Kolenda wrote,

> Basic to the *jajmani* system is commitment to a royal or lordly way of life as an ideal.[27]

The *jajmans* tried to live up to this ideal by surrounding themselves with as many followers as possible. Srinivas speaks of an "invest-

24. For instance, Wiser, *op. cit.*, 63, 95; Epstein.
25. Breman, 1969, 397.
26. Weber, English translation 1958, 105.
27. Kolenda, 1963, 21.

ment in people," an endeavor to put others under an obligation and thus maneuver them into a relation of dependence.[28]

Besides prestige, the *jajmans* also drew power from the number of their followers, and it was important to have power because of the rivalry existing among the patrons. The dominant caste was divided into competing factions, alliances of families, each with a following which could be mobilized owing to the claim of the *jajmans* to their *kamins*. Conversely, the fate of the protégés was closely bound up with the well-being of their protectors. A member of a low caste who was a follower of a prominant patron was assured of security and safety in times of scarcity and tension.

Wiser was right to the extent that members of the dominant caste were committed to render counterservices; but these were services in which their patronship was manifest. The clients were proud of the esteem enjoyed by their lord and participated in it. Lewis and Barnouw cite the following passage from the autobiography of a member of an impure caste:

> This is the only means of livelihood open to us, and the richer the landlord we serve, the more prestige and honor we have.[29]

The clients profited by the prominence of their patron to improve their own position as much as they could.

As Gould and others wrote, the *jajmani* system was an expression of the extremely hierarchical character of the caste structure. The element of patronage, on the other hand, cut through the horizontal stratification of the village community. If the dominant caste was internally divided and did not form a closed front, it was equally difficult for the *kamins*, even if they belonged to the same caste, to take a common stand. The hierarchical social structure was split up into blocs of patrons with their clients.

Patronage was an essential aspect of the *jajmani* system. But we should not associate the patron-client relationship exclusively with this system of exchange on the local level. Village patrons often were themselves clients of regional "overlords," for whom they performed various services and to whom they owed loyalty and support.[30]

28. Srinivas, 1955, 30; see also Beidelman, *op. cit.*, 27.
29. Lewis and Barnouw, 1967, 113.
30. The importance of a regional elite as a link between the societal suprastructure and the village community was pointed out by Frykenberg and others (Frykenberg, 1965).

A representation of *jajmani* relationships in terms of patronage implies that servitude in the past cannot be simply described as a system of unfree labor. The servant was not only a laborer but also a client, and as such he was entitled to affection, generosity, and intercession on the part of his master, who, as a patron, had to guard and promote the interests of his subordinate. Conversely, the servant owed his master respect and loyalty. His obligations were not restricted to labor, but included serving the interests of his patron and standing by him in all contingencies. Total accountability on the one hand and unconditional loyalty on the other exceeded the obligations of the delivery of an amount of labor and a suitable remuneration.

Does this description of the *jajmani* system as a relation of patronage imply, then, that Wiser's view was correct? It is true that the group contrasts ran vertically as well as horizontally through the caste structure. But it would be premature to conclude that intercaste relations were harmonious and afforded maximum satisfaction and security to all concerned. In the relationship of patronage, the unequal control of scarce goods was at the same time accentuated and corrected.

That the *jajman* as a patron strove to increase his power and prestige, and not primarily his income, made the relationship pattern no less unequal. The patron's own interest was all important, but to reach his ends he had to make sure, to some extent at least, of the affection and loyalty of his *kamins*, his clientele. The *jajmani* system as a form of patronage implicity mitigated exploitation, while at the same time it contributed to its continued existence.

It should be borne in mind that not all *kamins* were in the same position. The distance between them and their *jajmans*, as well as the character of the services rendered, differed widely. The relationship certainly tended to contain more elements of exploitation as a *kamin* fulfilled less favored, lower skilled, and less monopolized tasks, for instance, those in agricultural labor. An oversupply of precisely this hardly specialized category made the agricultural laborer easy to replace and thus increased his dependence.

It is, finally, of importance that the members of the artisan and serving castes worked for various *jajmans* at the same time, while the farm servants by the nature of their work were mostly bound to only one *jajman*. Though not essential, this difference did have some influence on the relationship of patronage for, unlike the farm servants, the other *kamins* had to distribute their loyalty among different *jajmans*.[31] Even when the latter belonged to the same faction within the dominant caste, the less attached *kamins* formed an element of uncertainty for the patrons mutually competing for power and prestige. The great ritual distance, the laborers' small economic importance as measured both by indispensability and by exchangeability, their continuous bondage—all these factors operated against the farm servant and detracted from the principle of reciprocity in his relation to his master. But even in such cases, complete exploitation and excessively arbitrary treatment were contrary to the interests of the landlord, who had to be sure of the support of his followers.

To conclude, servitude in the traditional community should be interpreted as a form of unfree labor, complicated and mitigated by a relationship of patronage. For a detailed documentation of this view, the reader is referred to chapter 4, which includes a description of the *hali* system in south Gujarat.

The changes in servitude should therefore be judged against the background of changes in the relations between *jajmans* and *kamins*. I have consistently spoken of the *jajmani* system in the past tense, because the intercaste relationships have changed so radically that, from our point of view, the use of this term in the present village situation is no longer justified.

In the past, fluctuations of any kind did not essentially affect the foundation of the system. We do not overestimate the stability of the *jajmani* system if we say that, for instance, political and administrative changes, economic developments and demographic shifts did not assume anything like the proportions of those that took place in the twentieth century. Nor did a completely closed village economy ever exist. It remains true, however, that the relationships lost their local flavor in a process of enlargement of scale. Commercialization of agriculture and continuously increasing government intervention diminished the importance of local autarky and autonomy. The drift of members of the artisan and serving castes to the urban centers, from which they began to serve the surrounding countryside, contributed to the rise of an ever-increasing number of different contacts which went beyond the village.[32]

31. Wiser, *op. cit.*, 164; Srinivas, 1955, 30f.; Beidelman, *op. cit.*, 46; Gould, 1964, 24.
32. Wiser, *op. cit.*, sect. III; Beidelman, *op. cit.*, ch. IV.

Depending on the accessibility of the region, this development began early or late. In the countryside of south Gujarat, outside influences had begun to increase sharply in all respects by the end of the last century. Because goods and services were supplied from urban centers—these were in part new goods and services, and paid for in money—the relationships which had existed on the local level among the members of the various castes no longer fitted the new situation and became partly or completely superfluous.

The changing structure of exchange was accompanied by changes in the norms and values on which the intercaste relationships had been founded. The element of patronage within the framework of the village gradually receded into the background. In other words, a process of instrumentalization contributed to a decline of the traditional rights and obligations on either side. Transactions were increasingly concluded on a money basis. *Jajmans* and *kamins* came to regard each other as employers and employees rather than as patrons and clients, and felt themselves bound more by contract than by status. While the mutual separateness more clearly manifested itself as the vertical attachment disappeared, the powerlessness and subjection of the lower castes received a sharper emphasis.

These very general and brief remarks will be elaborated and elucidated in my analysis of the change in the relationships between landlords and agricultural laborers. Here I only wish to add that this is the situation to which, in fact, Beidelman's view applies. The disintegration of the *jajmani* system was due, I believe, above all to the depatronization of the intercaste relationships as they had been institutionalized in a more or less closed local order. In Beidelman's study the time perspective is vague, but the data with which he elucidates his arguments are relatively recent. On the other hand, what Wiser depicted was, as he himself stated, the traditional village community—and in an ideal-typical sense—as it no longer existed at the time of his stay in Karimpur. A more dynamic interpretation of the intercaste relationships is therefore needed, that is, an interpretation which takes into account the shift in these relationships and the disappearance of the *jajmani* system.

The *jajmani* system as such will not be discussed further. The remainder of my study is concerned exclusively with landlords and agricultural laborers in south Gujarat, the Anavil Brahmans and the Dublas. The data which I collected on the relations among the other castes and on those between these castes and the Anavil Brahmans and Dublas were incidental.

In my view, the developments in the relations between these two agrarian categories constitute an important aspect of the larger changes that have taken place in the rural social and economic structure of this region since the nineteenth century. Hence my study at the same time throws some light on the nature of the changes in the relations among the other castes in this period.

PART II

Chapter 3

South Gujarat: Region and Population

3.1 Geographic Aspects and Historical Development

South Gujarat, in this study, is the name of the Indian region which administratively comprises two districts: Surat and Bulsar. Geographically a part of the western coastal plain of the Deccan plateau, it is an alluvial zone that lies more than 100 miles to the north of Bombay. This region can be divided into two parts. In the west is a densely populated and fertile plain composed of black clay and separated from the Arabian Sea by a narrow and infertile strip of saline soil. Toward the east, a barren area, with a comparatively sparse population and marked by an eroded and sloping landscape, gradually becomes hilly and wooded. In the south, on the boundary between Gujarat and Maharashtra, the black clay belt is almost wholly absent and the hills extend practically to the coast, whereas to the north, the plain broadens increasingly. The fertility of the clay region is surpassed only by that of the loamy soil on both sides of the rivers running from east to west and emptying into the Arabian Sea.

Despite some large irrigation projects completed during the last decade, agriculture is still largely dependent on the rains, which fall only during the monsoon, from the beginning of July to the end of September. Precipitation varies from about 40 inches in the west to 90 inches in the east. After the rainy season, winter extends from October to March, with temperatures averaging about 60° F. During the hot summer, from March to the end of June, temperatures rise to an average of nearly 100° F., although local differences are great.[1]

1. The *Surat District Gazetteer* of 1962, compiled on the model of the colonial *Gazetteers*, gives a detailed survey of the geography, history, and economy of the region, as well as an outline of the composition of the population. Various items of information in this chapter were taken from this work of reference.

The greater part of the population lives in the central plain and carries on intensive agriculture. The system of roads is largely concentrated in this area, and the most important urban settlements are near the coast along the railroad that connects Bombay with northern India.

In 1961, south Gujarat had almost 2,500,000 inhabitants. Eighty-five percent of these lived in the country, largely in villages of 500 to 1,500 inhabitants. The remaining 15 percent were in towns, that is, urban settlements of more than 5,000 inhabitants. In the central zone, the population is particularly dense, numbering upwards of 600 inhabitants per square mile—a figure exceeded in few other regions of the state.[2]

South Gujarat, an ancient culture area, is mentioned as a center of legendary kingdoms in the distant past. The more recent history of the region is closely bound up with the fortunes of the city of Surat, situated near the mouth of the Tapi River. In 1573, Akbar conquered the city, which under Moghul rule in the seventeenth century developed into the most important seaport and trading center on the west coast of India. During that period of prosperity, trading stations for resident factors were established in Surat by various European nations: there were Portuguese, English, Dutch, and French factories. Travel accounts of those days written by Europeans contain many details of daily life in the city and on the trade route to the north, but the region inland from the coast of this harbor principality is rarely mentioned. The rural part of south Gujarat belonged to the Moghul empire, but owing to the excentric situation of the region, central authority was always weak there. Reclamation of fresh areas by the founding of new villages was encouraged, and those who had already built up a local power domain were confirmed in their authority by Moghul overlords. As government functionaries—called *desais*—they were entitled to part of the revenues from the land tax. These frontier-region notables on the local and higher level successfully combined their intermediary function with a great deal of independence and recognized their subordination in name only.[3] The collapse of the Moghul empire brought with it a period of great unrest for the whole of south Gujarat, and a labile balance of power which lasted until the nine-

2. *Surat District Census Handbook*, 1962, 16.

3. The literature on the system of administration in south Gujarat during the era prior to colonial rule is extensive. Some of the main sources are Pedder, 1865; Watson, 1886; and Rogers, 1892.

teenth century. Historiographers have paid attention almost exclusively to events in the city of Surat, but the few data we have on the surrounding countryside show that there the administrative apparatus created by the Moghuls crumbled away fairly soon. Already in the second half of the seventeenth century the whole province suffered from raids by the Marathas, who wandered about ransacking the countryside. In the first half of the eighteenth century the Marathas conquered the whole region except Surat, which the Moghuls had leased to the English. Although the city still numbered over half a million inhabitants at the time, prosperity had ended, partly because the East India Company had moved its seat to Bombay.

The Polish traveler Hove, who made a journey southward from the city of Surat at the end of the eighteenth century, described a wooded and fairly inaccessible region. The population was concentrated in the coastal plain, and only there were they engaged in agriculture.[4] The power vacuum in this chaotic period was utilized by the local elite to enlarge its influence. In the absence of a regular administration its members acted as revenue farmers under Maratha rule. This country nobility came increasingly to regard the region in which they collected the taxes as their private domain, which they could dispose of at will. When they arbitrarily raised the tributes they levied from the population, the only form of protest available to the village inhabitants was migration, and this was widespread. One of the military leaders of the Marathas made himself independent of the ruling dynasty in Poona in the second half of the eighteenth century and founded the principality of Baroda. He then entered into an alliance with the East India Company, which enabled him to keep his territory in south Gujarat.

Then, after some campaigns at the beginning of the last century, the larger part of this province fell into the hands of the English. With this conquest, a long period of unrest ended. In the area under British rule the *desais* were displaced and, unlike their fellow caste members in northern India, they were not recognized as *zamindars*.[5] The system of revenue farming was replaced in 1817 by the so-called *ryatvari* system, under which each farmer was taxed individually. Settlement surveys in the second half of the nineteenth

4. Hove, n.d., 93–94.

5. In effect the situation did not change very much. At the end of the last century, Rogers wrote about the former power elite: "So thorough was the influence they originally obtained that up to the present day there are few villages in the southern districts of the Collectorate in which the . . . Patels are not descendants of the old Desais." (Rogers, 172.)

century served as a basis for the collection of land tax. The agricultural area of the plain gradually increased. Thanks to the *pax britannica,* the less fertile eastern part of south Gujarat became more accessible. The founding of villages in that region by members of high castes led to sedentarization of the local tribal population, which until then had led a migratory existence and had been engaged in primitive agriculture.

The construction of the railroad through the plain to northern India—the section between Bombay and Surat was finished in 1864—liberated the region from the relative isolation in which it had found itself after the fall of the seaport. With improved transportation, the production of cash crops was stimulated. In the north, cotton was introduced, a crop which thrived on the black soil. In the south, the cultivation of sugar cane was expanded. There was, however, no question of a sudden shift, no rapid transition from a subsistence economy to a market economy. Part of the produce had always been sold outside the village or transported to Surat. Moreover, although the construction of the railroad was an important advance, transportation of agrarian produce remained difficult as there was no system of roads connecting the villages with each other and with the railroad. Construction of such a network was not begun before the end of the last century, hence for a long time marketing continued along traditional lines. Until a few decades ago, urban traders bought up the crops at the farms and took care of the transport themselves. Early in this century the people of the villages still grew many different crops to provide as much as possible for their own needs. No rapid increase in the percentage of money income of the farmers occurred until the First World War.[6]

The same was true of industry, which after the decline of Surat had languished on in the small towns of the province, and which has grown rapidly only within the last few decades. After Independence, a number of new industries and workshops were started in urban centers along the railroad.

Finally, improved transportation has brought Bombay within easy reach, and a yearly migration to the brickyards and salt-pans near that city has begun. The migrants are small farmers and agricultural laborers for whom there is not enough work in agriculture during a large part of the year. As this rapid survey shows, a basis was laid during the nineteenth century for shifts in the economic

6. *Baroda Economic Development Committee Report,* 1918–19, 1920, 56.

structure which did not begin to take effect until the twentieth century.

Politically and administratively, too, colonial rule caused much less of a breach with the past than might be expected. In the part of south Gujarat that belonged to the principality of Baroda, the system of revenue farming continued to exist until the end of the nineteenth century, but the influence of the former authorities in the area under British rule also went on unimpaired. The lower functionaries were mainly recruited from their midst, and on the local level they could continue to exert their influence owing to the circumstance that, in the initial stages of colonial rule, decisions were largely made on the district level or below it. The far-reaching freedom of action enjoyed by the lower functionaries was curtailed only when a generally operative system of rules and procedures had been worked out and the powers of the various officials had been centralized and standardized.

In the principality of Baroda, government administration was increasingly adjusted to that of the neighboring English territory. Uniformity was completed when, a few years after Independence, the principality was incorporated into the State of Bombay. In 1964 the southern part of the Surat district, at that moment the largest in the State of Gujarat which had been formed in 1960, was split off as a new district with Bulsar as capital. Throughout this study the term of Surat district relates to this district in its old form.

3.2 Population

In the social structure, the most striking aspect is the high proportion of tribal elements in the total population. More important than the classification according to caste is the division of the population into *ujliparaj* and *kaliparaj* (light-colored and dark-colored), that is to say, into caste Hindus and tribal communities. The *adivasis* or "scheduled tribes" — as the tribal groups are officially called instead of the denigrating *kaliparaj* — constitute a little over 10 percent of the inhabitants in the whole of south Gujarat, but in the southern part of the state they account for half the total population.[7] But even this is an arbitrary classification, an incidental statement of facts. "The entire course of Indian history shows tribal elements being fused into a general society," wrote the historian Kosambi, and this

7. The latest *Surat District Census Handbook* (1962, 22) gives a percentage of 49.96.

completely applies to developments in south Gujarat.[8] In the eastern area, which until recently remained difficult of access, tribal groups have long been able to retain their independence. The most important among them are Chaudharis in the north and the Dhodias in the south. At the beginning of the last century they were shifting cultivators, who migrated when the roughly tilled grounds were exhausted and who returned to them only after many years. Pacification by the British made possible an accelerated penetration of caste Hindus from the plain; their coming contributed much to the opening up of this region and to sedentary agriculture.[9] When the colonial authorities introduced the *ryatvari* system they took account of the weak economic condition of the autochthonous population; they determined the amount of land tax to be paid not only according to the fertility of the soil, but also in the light of the social position of the cultivator: a lower tax was imposed on plots tilled by tribal cultivators. However, this well-intentioned measure soon led to transfer of property rights, especially those of the best plots, to newcomers who were attracted by the low tax. Nevertheless, caste Hindus did not pour into this barren, hilly region in great numbers, and in these eastern areas more than 80 percent of the population still belong to tribal groups. Most of them have come to depend as tenants on, for instance, Banias and Parsis, who settled in small market towns as money lenders and licensees for distilling and selling spirits.[10] In the course of the years a great many religiously oriented movements occurred here; they opposed the consumption of meat and alcohol, and in general propagated Hindu values. This is now done in a more institutionalized way by education in *ashrams,* centers of social action along Gandhian lines that have been founded in recent decades with the aim of emancipating the tribal population.

Whereas in the eastern regions the incorporation of tribal cultures into the economy and the social and religious system of the Hindu society is still going on, this process has come to an end in the plain of south Gujarat. Here a hierarchy with many different agrarian, servant, and artisan castes crystallized long ago. Several groups in this region, however, appear to be of tribal origin. The Kolis, for in-

8. Kosambi, 1956, 25.

9. With reference to the autochthonous population, the British district officer Bellasis wrote in the mid-nineteenth century, "They rudely till the soil, but are very migratory and unsettled cultivators. . . . They are experts in the use of the bow." (Bellasis, 1854, 3.)

10. For a more extensive survey of the social history of this region, see Baks, 1969, ch. 3.

stance, who are now included among the *ujliparaj*, were regarded as "aboriginals of the plains" in the early days of colonial rule.[11] The Dublas, on the other hand, who inhabit the plain, are still classified as a "scheduled tribe." All such groups, both those in the plain and those in the eastern regions, firmly consider themselves Hindus. In 1911, the Dubla pupils of one school sent a request to the government for permission to state their religion as Hindu in the census of that year. Up to that time, they had been classified as animists, even though it was already known in the nineteenth century that they kept the fasts and festivals of the Hindu calendar. During the campaign led by Gandhi for the abolition of untouchability, some Dublas refused any contact with those caste Hindus who had admitted the impure Dheds to their houses.[12] A Brahman priest establishes for Dublas the most favorable day for marriage and performs the required ceremonies. Their dead are now usually cremated, not buried, although the survivors can hardly afford the expense. In the mutual contact among *adivasis*, status is apparent from their willingness to accept boiled food and water from each other. Dublas, for instance, are placed lower in the scale than Dhodias.[13]

As a rule, the groups with a tribal background regard themselves as *rajputs* and try to substantiate their claim to this status by genealogical myths.[14] Members of high castes put much stress on the deviating norms and customs of *adivasis*, for instance in family life and religion, without nowadays meaning to exclude them from Hindu society. The term of *adivasi* is here synonymous with low social status; it implies a pattern of behavior which is inferior to that of the higher castes. "Tribal caste" seems the most suitable name for these groups, which are placed below the artisans and above the impure castes in the hierarchy.

In the plain of south Gujarat, the caste system prevails in the form in which it is usually described for village India. Artisan and serving castes—such as Goldsmiths, Carpenters, Potters, Tailors, Washermen, Barbers, and Shoemakers—live in the larger villages, almost always represented by more than one household. This also applies to the impure castes, such as Tanners and Scavengers. Mediator groups are especially Banias and Parsis, a growing number of

11. *Census of India*, 1941, vol. XVII, pt. 1, 62.
12. *Census of India*, 1921, vol. XVII, pt. 1, 134. Much ethnographic material on the Dublas is to be found in the study of Shah, 1958.
13. Solanky, 1955. See also *Census of India*, 1901, vol. XVIII, pt. 1, 502.
14. *Census of India*, 1931, vol. XIX, pt. 1, 460.

whom have left the countryside and settled in towns. The Parsis are not caste Hindus. The Moslems have in most cases always lived in the large and small towns of south Gujarat, where they earn their living as shopkeepers and artisans. By far the larger part of the population in the plain belongs to the agrarian castes, among whom Anavil Brahmans, Kanbis, Kolis, and Dublas are the most important.

The Anavils of south Gujarat do not fulfill any priestly functions but are peasant Brahmans.[15] Although they are not very numerous, their position as the chief landowners and local power elite leaves no doubt that their dominance in this region is traditional. They are mentioned as those who developed south Gujarat in a remote past by using tribal labor. The remark by Baines, which I shall cite in full in the next chapter, refers in part to this colonization by Brahmans:

> Wherever they have settled in large masses . . . they have taken to cultivation on the same lines as the ordinary peasantry, except that they but very rarely put their hands to the plough, though they go as far as standing upon the crossbar of the harrow to lend their weight to that operation. Owing to this caste-imposed restriction . . . he is the centre of a well-defined system of predial servitude, his land being cultivated for him by hereditary serfs.[16]

It is not clear whether there were any Anavils among the Brahman groups who came to south Gujarat. It is not impossible that they were already present as notables in the area, and went through a process of brahmanization to legitimize their secular supremacy religiously. The Anavil Brahmans are subdivided into two status groups, the Desais and the Bhathelas, the latter regarded as being lower. In the precolonial period the Anavils were included in the village and regional administration as persons of authority, and later as revenue farmers. Indeed, the administrative and political functions were so closely connected with this caste that the professional title of such a functionary became the name of a group within the caste. Transformed into a landed gentry, the Desais increasingly removed themselves from their fellow caste members who continued to lead a peasant existence in the village. The traditional distinction resulted in a relation of hypergamy between the two sections.[17] Un-

15. Enthoven, 1920, pt. 1, 223.
16. Baines, 1912, 28. See also *Census of India,* 1911, vol. XVI, pt. 1, 282.
17. This subject is treated in Van der Veen, 1972.

der colonial rule, however, the Desais were not recognized as persons in authority, and when revenue farming was abolished they lost their chief source of influence and prosperity. The Bhathelas, on the other hand, managed to improve their economic condition. The social decline of the Desais and the economic rise of the Bhathelas ushered in an equalizing process that began in the nineteenth century and is still continuing. The Anavil caste as a whole has remained the leading group in the province as a caste of landowners, and its members have utilized their dominance to build up a lead of the other castes outside the village community in the new urban professions as well.

In the northern plain of south Gujarat, outside the territory of the Anavils, the Kanbi Patels are the dominant element. From simple farmers with a presumably tribal background they gradually developed into a local elite, but they always remained village notables who were not charged with administration on the local level. Only since the beginning of the twentieth century has their influence increased, partly in consequence of their participation in the independence movement. In the center of their territory, Bardoli *taluka*, Gandhi and Sardar Patel led a successful campaign in 1928 for reduction of the land tax. Since Independence, the Kanbis have increasingly outrivaled the privileged Anavil Brahmans, both in the villages where they live together and in the urban economy.

Kolis form the largest caste of south Gujarat. Their territory roughly coincides with that of the Anavils, i.e., they live in the western *talukas*. The Kolis have always been chiefly small farmers, but also sharecroppers, tenants, and agricultural laborers for higher castes who own more land. Especially among the younger generation there are many who have found employment, in the last few decades, as factory hands and lower technicians within and outside the area. Economically they have progressed, and as the political importance of numerical ratios in the village community is growing, the Kolis may be regarded as an ascending caste.[18]

The numerical strength of the castes in south Gujarat is not known, but the proportion of Dublas in the total population was established when they were counted separately as a "scheduled tribe." They form about 10 percent of the total population of the region; in the center of their territory, the central plain, this per-

18. For data concerning the Kolis, see Hommes, 1970.

centage is much higher.[19] Gandhi called them *halpatis,* and this name has found increasing acceptance as a caste name. The earliest colonial reports already refer to them as the agricultural laborers of the high landowning castes, who employed them as bound servants. Most members of this tribal caste are still agricultural laborers in the rural areas; in 1961 this was the occupation of 78 percent of the members of the caste.[20] There are a few other castes of agricultural laborers in the plain, for instance the likewise tribal Naikas, but it is especially the much more numerous Dublas who are looked upon as such. The population of agricultural laborers is strongly represented also in *talukas* with a high Dubla concentration. Outside the territory of the Dublas, notably in the eastern *talukas,* the percentage of tenants among the agrarian population rises sharply, as was shown in the 1951 census returns.[21] This is to be understood primarily as a different organization of agrarian production. It is doubtful whether Dublas in the fertile plain ever have owned any land on a large scale as independent farmers. We cannot even guess when, but it was certainly long ago that they became dependent as agricultural laborers. In the eastern area of south Gujarat, on the other hand, a relatively small number of incurrent members of high castes managed not very long ago to gain control of a large part of the arable land, which they subsequently leased to the original owners. In this area, too, a relation of dependence arose, but one of a different nature.

The low placement of the Dublas in the caste hierarchy parallels their weak economic position. Despite age-old contacts with landlords of higher caste, numerous tribal elements can be discerned in the social organization and culture of the Dublas. Their low living standard has probably prevented them from being sanskritized more rapidly. Recently, however, there has developed a tendency among the Dublas to abstain from meat and alcohol, to take morning baths, and to observe Hindu festivals.[22] Such attempts to conform to the behavior of higher social strata partly express the Dublas' protest against their dependent and less favored position in the social system.

Formerly, the members of the various agrarian, serving, and artisan castes in the plain of south Gujarat mutually maintained *jaj-*

19. *Surat District Census Handbook,* 1962, 200.
20. *Ibid.*
21. *Surat District Census Handbook,* 1952, 12–15.
22. Joshi, 1966, 91–93.

mani relations. In the network of exchange relationships, locally called *avat*, the Anavil Brahmans occupied a central position as the most important landowners and local power elite. "The social economy of the area is thus organized around the Anavil land owners, almost all the other castes being in functional dependence on them," as Naik wrote. Joshi arrived at the same conclusion: "It may be said that it was the group of Anavils on whom, more or less, depended the other caste-groups of the village for their earnings." [23]

Mukhtyar describes the founding of a village as it took place more than 200 years ago. The Anavils attracted members of agricultural, artisan and serving castes to work for them in exchange for a grain allowance or a piece of land.[24] In his monograph on a village in south Gujarat, Joshi makes it clear that agricultural laborers formed part of this system of exchange:

> As the Anavil masters supplied the necessary clothing to their Dubla-Naika Halis, the Darji's patrons were the village Anavils. Similarly the Mochi supplied shoes to the village people. As in the case of the Darji, the Mochi's patrons were the village Anavils because the latter provided shoes also to their Dubla-Naika Halis.[25]

In the remainder of this study I shall confine myself to the relations between landlords and agricultural laborers past and present. To conclude the above brief sketch of the area and the population, these two categories are not difficult to localize socially and geographically. They are primarily the Anavil Brahmans and the Dublas in the plain of south Gujarat.

23. Naik, 1958, 389; Joshi, *op. cit.*, 23. See also I. P. Desai, 1964, 83, 89, 160.
24. Mukhtyar, 1930, 44–45.
25. Joshi, *op. cit.*, 23.

Chapter 4

The *Hali* System in South Gujarat

4.1 Masters and Servants

The Permanence of the Service Relationship. In the earliest reports of British administrators, during the first decades of the nineteenth century, there was mention of a system of attachment—*hali-pratha*—which later became known as a form of bondage.[1]

Hali was the term applied to a farm servant who, with his family, was in the permanent employ of a landlord, a *dhaniamo*. The service was not contracted for a definite period; as a rule it continued indefinitely. It usually began when an agricultural laborer wished to marry and found a master who would pay for the marriage. The debt thus incurred attached the servant to the master for life. It increased in the course of the years, so that repayment was practically impossible. The service relationship ended only when the *hali* was taken over by another master. Although in such a case the original master received compensation, the transfer was not a business transaction; in several reports it was stated emphatically that *halis* were never sold.

Not only was entering into bondage the beginning of a lifelong service relationship, but the unfree elements in it were reinforced by some hereditary features. It is true that the debt incurred by the *hali* during his lifetime could not be passed on to the next generation, but the master had a claim to the son of his servant, as it was thanks to the master's beneficence that the boy had grown up, living on the *dhaniamo*'s food.[2] When the son married, usually at the age of sixteen, it was therefore natural that he should enter the employ of his father's master. Only when the master did not need another

1. *Correspondence between the Directors of the East India Company and the Company's Government in India on the subject of slavery. Account and Papers*, vol. 16, session 15–11–1837 — 16–8–1838, pt. LI, 433.

2. *Report of the Congress Agrarian Reforms Committee*, 1951, 129.

servant was the boy free to look for another master. These hereditary features were so essential to the relationship that, according to some reports, only those agricultural laborers who had been attached to the same family from father to son were called *halis*.[3]

The Social Background of Masters and Servants. As elsewhere in India, the servants came from castes of tribal origin. Most of the *halis* in the plain of south Gujarat were Dublas. In some regions the category of farm servants included Kolis, the poorest members of this caste of small farmers and sharecroppers, who, it was said, had fallen so low because of excessive drinking. They formed a separate group (*gulam* Kolis, i.e., slave Kolis), who had little contact with their fellow caste members. It was said that they were obliged to practice endogamy or even found partners among the Dublas. Members of other tribal groups were also mentioned as *halis*, but bondage has been identified with the Dublas to such an extent that for the members of this caste of agricultural laborers the term *halpatis* has become general. It is a euphemism for *hali*, introduced by Gandhi when he campaigned for the abolishment of *halipratha* in the nineteen-thirties.

Most *halis* were Dublas, but not all Dublas were permanently attached to a master. Lack of data makes it impossible to determine proportions; we can only say with certainty that bondage was the prevailing service relationship. In 1825, the Collector wrote that in Surat there were more "slaves" than anywhere else in India, and in one of the settlement reports of the end of the last century it was stated that the majority of agricultural laborers consisted of dependent farm servants.[4] There was, however, no clear-cut difference between bound and free agricultural laborers. The *chuta halis* (literally: free *halis*) belonged to the latter category. They, too, were indebted to a landowner, though not as deeply as the *bhandela halis*, the bonded servants. Unlike the latter, they did not work for the same farmer all the year round, but only when he needed them, and at other times they had to provide for themselves. The difference is not always very clear, but what it amounts to is that the *chuta halis* were less bound and, on the other hand, had less security. In

3. *Gazetteer of the Bombay Presidency*, 1877, pt. 2, 198.

4. "Some of these agricultural labourers are free to work wherever they like, but the majority of them are Hallee, either Bandhella or Chhuta," says the first survey and settlement report from 1893 of the subdistrict of Palsana, which is included in the *Revision Settlement Report Palsana Taluka*, 1911.

the sources I consulted there is no mention of a category of completely free labor.

The servants were invariably in the employ of members of prominent castes. These included Rajputs and Borahs, who were not very numerous but in some villages were the most important landlords. In the northern part of the plain, *halis* worked for Kanbi landowners, but it was preeminently the Anavil Brahmans who employed *halis*. They were the dominant caste in south Gujarat and, as such, the owners of the best and largest plots of land. The *hali* system, in other words, is especially identified with these two agrarian castes.

> The Dublas are the most degraded of all the agricultural classes. They were, for the most part, the "Halis" or slaves of the Anawala brahmins by whom they were employed in cultivation of their lands. Almost all families of the Dublas had been slaves for many generations.[5]

It would not be correct to represent the landlords as large landowners. Only a small portion of the Anavil Brahmans possessed so much land that they could rightly be called *zamindars*. These *zamindars*, the most powerful and prominent of the Anavil Brahmans, had many dozens of servants in their employ, and the same was true of some large landlords in villages where sugar cane, a labor-intensive crop, was grown. In contrast, most of the members of the dominant caste had only one or two servants. The least prosperous among them even had to do without any help whatsoever.

The Origin of the Hali System. "The Desaees Buttela Brahmins in Parchol, Soopa and some other pergunnahs, possess as large a portion of slaves as may be found perhaps in any part of India," wrote the Collector in reply to some questions on the incidence of slavery in his district shortly after Surat had come into the hands of the British.[6] Such clear evidence of the existence of the *hali* system even in the early nineteenth century refutes the view of Patel that this form of bondage arose under and in consequence of colonial rule.[7] As he mentions the *hali* system explicitly, not much is left of his argument as tested by the situation in south Gujarat.

Kosambi, and others, connected the rise of bondage with periods of famine. It was thought that in consequence of such a crisis, tribal groups became enmeshed in a debt relationship from which neither

5. Choksey, 1968, 55.
6. Letter from W. J. Lumsden, Collector and Magistrate of Surat, dated 8-8-1825 and included in the correspondence referred to in fn. 1 above.
7. Cf. my rejection of his argumentation in chapter 1.

they nor the following generations could escape.[8] As an explanation, however, famine seems too incidental an occurrence to give rise to an institution of such importance, which governed the relations between landlords and agricultural laborers in south Gujarat for so long a period. Instead of attributing decisive importance to events that took place in the historic past, it is more fruitful to retrace the general circumstances that favored the *hali* system.

In this connection, it is significant that the Anavil Brahmans are regarded as those who opened up the plain of south Gujarat. These Anavils are a caste of "*grahaste*" Brahmans, that is, those who are not allowed to perform any priestly functions. Concerning these landowning Brahmans, who also live elsewhere in India, Baine writes,

> Wherever they have settled in large masses, as in the Gangetic Doab or Oudh, or in compact local colonies, which probably preceded their advance as a sacerdotal body, they have taken to cultivation on the same lines as the ordinary peasantry, except that they but very rarely put their hands to the plough, though they go as far as standing upon the crossbar of the harrow to lend their weight to that operation. Owing to this caste-imposed restriction, probably, it may be noted that where the Brahman has settled otherwise than as a part of a large general community, he is the centre of a well-defined system of predial servitude, his land being cultivated for him by hereditary serfs of undoubtedly Dasyu descent. This is the case of the Masthan of Orissa and Gujarat [i.e., the Anavils], and with the Haiga or Halvika of Kanara, and the Nambutiri of the Malabar Coast, all of whom have settled in very fertile country.[9]

Various local legends, also in south Gujarat, recount how *Rama* distributed the land among colonizing Brahmans and provided them with tribal labor to reclaim and till this land. As in the case of the other peasant Brahmans, it is not certain whether the Anavils came from elsewhere. At any rate, there is no doubt that agricultural groups of this kind managed to lay claim to the land in their present territory at an early stage. Whether in doing so they dispossessed any tribal communities, particularly Dublas, is not clear. Presumably the latter, who were hunters and shifting cultivators, gradually—and especially in periods of scarcity—came to depend upon a class of substantial peasants, whether or not of alien origin, who had been hinduized at an earlier period, and who ultimately found recogni-

8. Kosambi, 1956, 353.
9. Baines, 1912, 28.

tion as Brahmans. The economic and social subjection of the tribal population in the plain may have laid the foundation for the Anavils' brahmanization.

This process cannot be dated with any certainty, but it probably took place long before the time of the Moghuls. As there are no historical data to confirm such a development as is sketched above, only the plausibility of the explanation can be demonstrated.

LABOR CONDITIONS. *Hali* means literally, "he who handles the plow (*hal*)." It is not an arbitrary designation, but indicates an activity which members of high castes in many regions of India try to avoid because of its impure character. The *hali* bound not only himself, but also his wife and children. The farm servant worked chiefly on the land, but he was also the personal servant of the landlord—an odd-job man who attended him, ran errands for him, in short, assisted his master in everything and did what could reasonably be expected from him.[10] Neither the work nor the working-hours of the *hali* were clearly defined. At all times of the day, and if necessary of the night, he was at his master's beck and call, always ready to carry out orders.

His wife, the *harekwali,* served as maid in the house of the master, usually only in the mornings. Her tasks ranged from grinding grain, fetching water, sweeping the floor, washing up after meals, and cleaning the stable to emptying the chamber pot on the dunghill. She did the rough household work which elsewhere was done by members of the impure castes. Moreover, she helped out on the land in the busy season.

The servant's daughter assisted her mother in the household of the master. His son—as *govalio* (cowherd)—tended the cattle and did various light jobs. In this period the master could test him to see whether he qualified both in ability and in personal behavior for later employment as a servant.

10. At the beginning of the nineteenth century the duties of a bound farm servant in Bihar were described as follows: "The master might employ slaves in baking, cooking, dyeing and washing clothes; as agents in mercantile transactions; in attending cattle, tillage, or cultivation; as carpenters, iron mongers, goldsmiths, weavers, shoe-makers, boatmen, twisters of silk, water-drawers, ferriers, bricklayers and the like. He may hire them out on service in any of the above capacities; he may also employ them himself, or for the use of his family in other duties of a domestic nature, such as in fetching water for washing, anointing his body with oil, rubbing his feet, or attending his person while dressing, and in guarding the door of his house, etc. He may also have connection with his legal female slave, provided she is arrived at the age of maturity and the master has not previously given her in marriage to another." Cited in Lorenzo, 1943, 61.

Attachment for an indeterminate period, severance of the relationship only in exceptional cases and often its prolongation into following generations, work obligations for the servant's whole family, and finally the nonspecific and exchangeable nature of the service were the chief elements of servitude in the past.

The counterservice expected from the master could be summarized in the obligation to provide for his servants. As one of the first reports on the *hali* system commented, "rearing and feeding in years of scarcity and the charges of settling them in marriage (about 30 rupees) are all borne by the master."[11] On the days that the *hali* worked, he was entitled to an allowance of grain—the *bhata*—which varied according to the size of his family. Instead of this grain he sometimes received a piece of land to till with his master's land; he was allowed to keep the produce for himself. In addition, he was given a slight breakfast in the master's house, and in the busy season also a midday meal on the land.

The master allotted him a site where he could build his hut, and supplied the materials for it. Each year the *hali* received some clothing: a headcloth, a jacket, a loincloth, a scarf for the winter, and a pair of shoes. His wife was entitled to a *sari* and some brass ornaments every year, and a meal daily after work. Further, the *hali* was allowed to gather firewood on the master's land. When the day's labor was over he received some tobacco and sometimes some toddy (fermented coconut milk). When he was ill, the master provided medicines, and if he lived to old age, the master gave him food every day when he could no longer work.[12] As is evidenced by the extent of the benefits enjoyed by the *hali*, he received total care, that is to say, he depended for his every need on what the master put into his hands. Practically everything he received was in kind. Only on special occasions, decided by the master and granted as a favor, did the servant get a small amount of money instead of his grain allowance. He received some coins as drink money (*pivanum*) especially at the high religious festivals of *divali* and *holi*, and on festive occasions in the master's house. By the end of the nineteenth century, money had become a somewhat more important component of the total package of the master's counterservices, but for the year as a whole it remained a trifling amount. Most of it was spent on spirits bought from the Bania or Parsi in the village.

11. Lumsden, *op. cit.*
12. *Gazetteer of the Bombay Presidency*, 1877, pt. 2, 198–200.

In this connection it should be noted that the debt of the *hali*, though recorded in money terms, in fact largely represented gifts in kind. For the marriage feast of his servant the master supplied great quantities of food and drink, and, in addition, some pieces of clothing for the *hali* and his wife. As for ready money, the servant was given the amount that he had to hand over to the bride's father, rarely more than 10 rupees. The money value of the total expenses with which bondage began in the last century did not exceed 100 rupees, and usually it was far less. This, however, was only the initial debt, which increased as the years went by.

The master did not have employment for his servants throughout the year, certainly not for all of them. When there was little or no work for a *hali* he was allowed to work for someone else and keep his earnings. Yet it was precisely in the slack period that other landowners needed little extra labor. At first the *hali* lived on the credit he had saved up, withdrawing only so much grain from his daily allowance as he really needed for food. The rest remained to his account, but he could never save enough to bridge the whole of the slack period. When his reserve was exhausted—and it did not last long, the master being the only one who kept book—he was allotted a quantity of grain, an advance payment which was now called *khavati* (literally, to eat). At the end of the rainy season the account was made up and the balance he owed was added to his total debt. Illness and other accidental circumstances caused the servant to be ever more deeply indebted to his master, and repayment was obviously out of the question.

4.2 Servitude as a Labor Relationship

In the literature, the bondage of the Dublas in the past has been explained on purely economic grounds, and stress was laid on the fact that a number of landowners needed to have an adequate amount of labor permanently at their disposal. As stated above, it was the large landowners (chiefly Anavil Brahmans who did little or no work on the land themselves) who had one or more *halis* in their employ, as they were the only ones who could afford such assistance. The nature of the crops was another factor. In regions with an intensive crop system, especially where sugar cane was grown, there appear to have been more *halis* than elsewhere. They, however, had no year-round work either.[13] A motive at least equally important for

13. M. B. Desai, 1948, ch. 6.

the landlord was his desire to assure his having enough help at the peaks of the agrarian cycle. As agriculture depends on the rains, it is essential that certain activities, for instance sowing, transplanting, and harvesting rice, and plowing the land before a new crop is sown, should be finished within a few days. The normally abundant supply of labor very soon turns into a severe shortage. So as to have enough assistance at the most decisive moments, it was necessary for the landlords to take on permanently at least one farm servant, and if possible more, thereby monopolizing their labor and that of their families.[14] It should be added that to employ servants was not very expensive at the time.

The motivation of the landowning Anavil Brahman was as described above. Yet if it was advantageous for him to appoint Dublas in permanent service, what made the Dublas offer themselves as farm servants on such unfavorable terms?

That question is easy to answer when it is assumed that bondage was simply imposed. This is how the *hali* system has usually been represented. Great emphasis has been laid on the coercion applied by the masters, both in contracting and in maintaining the relationship. As already mentioned, it was not possible for the servant to abrogate the agreement, and his son was exposed to great pressure to follow in his father's footsteps. The servant was not allowed to leave his master until the debt had been paid off. It is therefore not surprising that the first British government officials in the early nineteenth century spoke of the *halis* as agrarian slaves, bound servants, praedial laborers, etc., a view which still has not essentially changed.

Unfree elements were certainly inherent in the *hali* system. But as pointed out in chapter 2, an interpretation which takes nothing else into account is too one-sided. For it is difficult to maintain that the servitude of the Dubla agricultural laborers was forced upon them against their will. On the contrary, for lack of continuous employment in the traditional agricultural economy, those who had managed to find someone to provide for them had every reason to consider themselves lucky.

It follows that servitude was in fact preferred to free labor by

14. "At first sight, this arrangement scarcely appears an economical one to the masters, but they have assured me that, everything considered, it is financially better than keeping no farm servants, and engaging labourers whenever they are wanted. For, when any of the principal operations, such as ploughing, reaping, etc., are in full swing, the demand for labourers is often far greater than the supply and not to have sufficient hands at such seasons must mean a heavy loss." *Papers Relating to the Revision Survey Settlement of the Bardoli Taluka*, 1897, 7.

landowners and agricultural laborers alike, and paradoxically enough for the same reason: the unequal distribution of the work on the land over the different seasons. Economically, the *hali* system was an attractive proposition for both parties. The Anavils wished to offset the risk of temporary labor shortage as much as they could, while the Dublas were well aware that there was work for them only during a limited period. As farm servants they did not receive any remuneration, but they were entitled to credit. Under the circumstances it was hardly relevant that their debt increased in this way. More important was the fact that their subsistence was assured in an economy of scarcity.

> The reason why he [the *hali*] sticks to the Dhaniama is that he helps him on all occasions of death, marriage, etc. In a way the Dhaniama is his sahukar [moneylender]. I called a few Hallees of each kind and questioned them closely and to my great surprise found that they were very contented with their lot.[15]

More telling than these testimonials of satisfaction before a British official are the statements in various reports from the last century that the *halis* were better dressed and had more and better food than those Dublas who were not attached.[16]

Dhaniamo means "he who gives riches."[17] At first glance such an appellation may seem somewhat exaggerated. The master's obligations did not go beyond providing the most indispensable means of existence for his *halis*. In the traditional agricultural economy, however, characterized as it was by widely varying crop yields and a strongly fluctuating labor demand, to be attached or unattached meant, in the final issue, either to have secured one's existence—however minimal—or to be so impoverished that a period of scarcity might easily bring starvation.

It is therefore more than doubtful that the *hali* strove to end his attachment. His being coerced to work is usually inferred from the condition that the servant was not allowed to leave his master as long as he was indebted to him. But the debt was rather fictitious in character and, if only for this reason, the term of debt slavery ap-

15. *Revision Settlement Report Palsana Taluka*, 1911, 5.

16. "I believe the slaves to be more comfortable than the free portion of their respective castes," wrote Lumsden in his letter of 1825. Half a century later this statement is unequivocally repeated: "*Halis* are still, as a rule, better off than those of their clan who are nominally free labourers." (*Gazetteer of the Bombay Presidency*, 1877, pt. 2, 201.)

17. B. H. Mehta, 1934, 339.

plied to this form of servitude is not very felicitous. Not only was repayment merely theoretical on account of the *hali's* minimal remuneration, but it was not envisaged by either of the parties.

For the master the expense involved in the beginning of the service relationship was an investment he had to make in order to obtain the services of a farm servant. He did, however, try to keep this debt within reasonable limits, as he well knew he could not recall it. The *hali*, on the other hand, did his best to maximalize it, and tried to get something out of his master as often as he could. That his dependence thus increased did not trouble him in the least. In view of his slender chances of finding work as an unattached laborer, repayment of his debt in order to end his bondage was the last thing he wanted. In short, servitude was sought rather than avoided by the Dublas.

So much was labor compulsion regarded as fundamental to bondage that some authors found it surprising that the *halis* were in no way inferior to the other Dublas.[18] Not only were they not held in contempt, a Dubla even was gratified if his daughter married one. Only then could he be sure she was provided for.

The option held by a master on the son of his farm servant also applied the other way around. Continuance of the tie from generation to generation was contingent upon the approval of the *hali* and his descendants. For them it was as much a right as a duty to succeed their fathers as farm servants.[19] The initiative to begin the relation was taken by both parties, and it was in the interest of both to maintain it.

Such an interpretation points in the direction of voluntary servitude. The motive of the master aside, the Dubla himself preferred permanent attachment, but it should be added that in general agricultural laborers chose dependence only to hold their own and to secure a livelihood. Nevertheless, it did not in any way alter their lack of freedom.

In both of the explanations of the *hali* system discussed above, the labor motive was the only one taken into account. In interpreting

18. "No social degradation attaches to the position of a Hali. Men of this class intermarry with the independent labourers of their own tribe." (*Gazetteer of the Bombay Presidency*, 1877, pt. 2, 198.)

19. Cf. also Harper, who describes a similar relation between landlords and agricultural laborers in Mysore: "Indentureship was a method of binding an employer and an employee into a stable alliance which ideally persisted from generation to generation. The servant was the lifelong employee of the master, and the servant's sons and the master's sons should, if possible, continue the alliance." (Harper, 1967, 40.)

this system, however, we should consider both the economic use of farm servants and also their function in the value system obtaining for landlords of the past.

To put it more concretely, *dhaniamo* and *hali* were not simply employer and employee. As explained in chapter 2, this was one aspect of the relationship; some elements of patronage could, however, also be distinguished. I shall examine this in the following paragraphs, beginning with the position of the dominant Anavil caste during the last century.

4.3 The Position of the Anavil Brahmans in the Rural Structure

It is the life style of a landowner that determines, in part, whether he needs any laborers and how he employs them. That is to say, labor demand is determined by considerations of a social nature which entail economic consequences.

A landlord's need of the services of a hali was largely inspired by his desire to work on the land as little as he could, to be exempt from activities which he thought disagreeable and in any case undignified. In other words, a *hali* was taken on primarily to replace the labor of the master and the members of his household, not to enlarge the total effort by his contribution.

As landlords (which is the point) it is the Anavil Brahmans who have become widely known, especially the category of the Desais, of whom the most prominent, the Pedivalas, resided in the small towns of south Gujarat and enjoyed great esteem as local representatives of Moghul rulers. The term of landed nobility applies to them insofar as their income was derived from land and from land taxes.[20] They formed a regional elite and were no longer physically concerned in agriculture. Illustrative of the impressive style in which these most prominent members of the dominant caste lived is a report of 1863 on the outcome of a conflict between the prince of Baroda and the *desai* of the subdistrict of Navsari. Thanks to the intercession of the British administration, the *desai*, deprived of his rights as a revenue farmer, received in compensation almost 200,000 rupees and—a telling detail—annual payments of 300 and 1,000 rupees to allow him to keep a carriage and a palanquin.[21]

20. "The failure to define satisfactorily just what is rural or urban has led some social scientists erroneously to assume that because many in the upper class are large landowners, the elite are therefore largely rural. Others, however, have recognized the fundamentally urban nature of this stratum." (Sjoberg, 1965, 112.)

21. R. Wallace, 1863, 468. T. B. Naik, 1957, 178, also underlines the prom-

Less prominent, but still clearly landlords, were the Desais who acted as revenue farmers on a lower level—in one or two villages—and who had acquired the administrative powers conferred by this appointment.[22] For them, the regional elite was a reference group, with whom they were also allied through marriage ties. The landholdings of the village Desais never attained the size one would expect in a landed gentry, not the least reason for which was the subdivision of their property among all the sons. In their quality of landlords these Desais remained active in agriculture, even if they confined themselves to supervision. They left the physical labor to their *halis*.

The great majority of the Anavil Brahmans belonged to the lower Bhathela category. These, too, were above the average as landholders, but they lacked the opportunities open to the Desais to augment their property legally or illegally as members of the local power elite. In every way these peasant Anavils were inferior to their Desai fellow caste members who controlled the villages and in the precolonial period acted as representatives and confidential advisers to the regional elite. Wheras the indolent Desais, in their style of life, reminded the British government officials of the *zamindars* in the north of India, the Bhathelas were praised by them as industrious and conscientious farmers.[23]

In the early nineteenth century a great distance separated the two categories, but they were not strictly held apart. It was always possible to pass from the Bhathela group into the Desai group and, similarly, members of the latter could rise to greater eminence within their own status group.

The rights and emoluments accompanying the quality of revenue

inence of the Desais in the past: "Apart from the monetary and village grants some Desais had their own insignia: a cannon, a *phalki*, a canopy and a *chamar*, and had their own torch-bearers and attendants. Moreover they had the rights of being gifted in kind or cash by various other communities under customary rules."

22. This appears, for instance, from *Morrison's Report* of 13–11–1812, in which the system of tax collecting is outlined. "The zemindars (Desais and Patels) generally in the end agree to two or more persons and authorize them to settle the *juma* payable by the whole purgunnah and afterwards to distribute the amount on the different villages."

The district officers, as they were called by the British government, stood surety for each other. Each subdistrict had a number of principal *desais*, but the rights were subdivided among many others. In Bulsar, for instance, there were five "principal" *desais*, but 140 "subordinate sharers." For other subdistricts, such as Chorasi, Chikhli, Parchol, Supa, similar data are given. See also fn. 24, below.

23. *Gazetteer of the Bombay Presidency*, 1877, pt. 2, 192. Pedder states that, at the beginning of the nineteenth century, the Desais were regarded as *zamindars* and also called themselves that. (Pedder, 1865, 316.)

farmer were, in south Gujarat, traditionally divisible and transmissible. The members of the family all participated equally in the influence and prosperity which the office of *desai* brought with it.[24] Moreover, that office could be transferred from the family by sale or sublease. In practice, the delegation or sale of rights was almost always restricted to fellow caste members—in other words, the office remained especially a prerogative of the dominant Anavils.[25] In the villages under their jurisdiction the high Desais appointed fellow caste members as deputies who had to collect the taxes for them, to keep order and to dispense justice. It was in this way that the more prosperous and influential Bhathelas, utilizing the coupling of function to position within the caste, graduated to the Desai category.

The introduction of a number of Anavils into the local administration, therefore, did not lead to final partition. Power and riches were not concentrated within a limited and closed group, but the benefits of *desai*-ship always remained open to individual Bhathelas who, when they succeeded, ultimately found recognition as Desais.

Apart from this, mobility was institutionalized in another way, i.e., in the relation of hypergamy between Desais and Bhathelas. A Bhathela was prepared to pay a large bride-price to marry his daughter to a member of the higher status group.[26] In this way the girl could look forward to a life of comfort. No longer would she be obliged to do peasant work, and in the household she would be assisted by servants. These considerations were not primarily inspired by fatherly care and affection: marriage was a family affair rather than a personal matter, and could place the Bhathela on a good footing with his prominent and influential fellow caste members. Such contact with Desais in itself already gave prestige and

24. "Among the Desais of the Surat Athavisi there seems to have long existed, as there still exists, a jealousy of each other which would not permit any individuals to assume the position of principals or representatives of the family in watan duties; all sharers were considered equal, and as the watan families increased, the members of them divided among themselves the Desais' rights in different villages of their Purguna, and indeed the Desaigari of the same village was often divided into the 'khatas' (i.e., part) of two or more Desais, brothers or cousins." (Pedder, 1865, 316.)

25. Pedder states that Bulsar, for instance, was divided into four Desai watans, but the attendant emoluments benefited 350 separate shareholders belonging to seventeen different families, "whose rights have mostly originated in purchase or mortgage." (Pedder, 1865, 313.) In the principality of Baroda the office of *desai* was still sold until the end of the last century. (*Baroda Enquiry Commission*, 1873–75, pt. 1, 519.)

26. Concerning this marriage system, see Van der Veen, 1972, 75–82.

enabled the Bhathela to claim, if necessary, intercession and aid from the family to which he had married his daughter.[27] The connections of his new relatives, finally, improved the Bhathela's chances of acquiring a *desai*-ship for himself.

It is not impossible that stabilization of the Desais' position under colonial rule would have led ultimately to complete prohibition of intermarriage between the two categories. There was no progressive differentiation, however, and in the last century a leveling process took place instead. It was precisely the great power and autonomy of the Desais that induced the colonial administrators not to make use of them as intermediaries, so that they lost much of their influence and wealth. Only with great difficulty did they adjust to the altered circumstances, and in many cases they went into debt.[28] Under the circumstances, hypergamy was actually a boon. Gifts from their Bhathela relatives by marriage made continuation of their often extravagant way of life possible.

The following moralizing quotation expresses very well the behavior of the Desais and their attitude towards the Bhathelas:

> The Bhathella Desais, owing to their influence and cunning have succeeded hitherto in keeping the Government demand to a disproportionately low figure, which enables them to take less personal interest in the cultivation of the land. Hence all the fine masonry wells in Eroo are out of order, and the Dab, a weed which may be aptly called the Nemesis of improvident tillage, holds up its head to reprove the Desai, who thinks it beneath him to be a farmer or a landlord in the sense the Kaira Desai is. His pride takes another form, the pride of caste and privilege to marry 3 and 4 wives from the cultivating Bhatella class, who give him from 2000 to 4000 Rs to effect a settlement for their daughters in his home.[29]

After much squabbling, the British administration finally made a new arrangement for the Desais. On the advice of an enquiry committee, which made its report in 1865, the descendants of the former power elite received an annuity in compensation for the rights of

27. See *Census of India*, 1911, pt. 16, vol. 1, 282, and Naik, 1957, 178.

28. See the first cadastral report of Bulsar *taluka* from 1870, included as appendix R in *Papers Relating to the Revision Survey Settlement of the Bulsar Taluka of the Surat Colectorate*, 1900, 70. Cf. also the correspondence between the Collector and other government functionaries, which is included in *Papers Relating . . . Jalalpor Taluka*, 1900, 84, 99. See also the autobiography of G. G. Desai, an Anavil who in the colonial bureaucracy rose to be a magistrate but was discharged with ignominy. (G. G. Desai, 1906, 3.)

29. Cited from the first cadastral report of Jalalpor *taluka* in 1868 and included in *Papers Relating . . . Jalalpor Taluka*, 1900, appendix R, 11.

which they had been deprived. These annuities, called *desaigiri* after their former remuneration, were paid until 1953. The amounts were small, the more so as they had to be distributed among the ever-growing families. More important than the money, however, was the fact that the Desais could thereby legitimize and strengthen their claim to former importance.[30] Documents to show the rights formerly granted are still preserved with care. The solemnity and ostentation with which until recently the descendants of the *desais* went to town on the annual payday to receive their annuity had no bearing on the few rupees to which they were entitled, but was meant as a pointed reminder to the other village inhabitants of the illustrious past of their families.

A background which made it clear that a family had politically influential ancestors remained a prestigious factor until this century, as did residence in a village traditionally known as a domain of the higher status group. But from the early nineteenth century onwards—fairly soon in the British part, later also in the principality of Baroda—the Desais were compelled to live exclusively on their property in land. They did this in accordance with their aristocratic ideas, that is, as landlords, and looked down contemptuously on those of their fellow caste members who, even though they employed a servant, had to work with him in the field.

4.4 The Anavil Brahman as a Patron

A landlord's style of life, not the function of *desai*, had become the most important standard of evaluation for the Anavils. Having one or more servants at one's disposal indicated a certain degree of prosperity and enabled the landowners to behave in a manner which rated high in the scale of values and which, though on a more modest footing and carrying less power, in fact guaranteed for this dominant group continuity with the past.

30. "With regard to the future entry of names, the Collector may guarantee to the *Watandars* that a record shall be kept in a *Watan* book to show the shares of each sharer whose name is now entered, and that this book shall be corrected from time to time by the insertion of the names of heirs in application and legal proof of heirship. But no guarantee of payment of separate small sums in the form now paid should be given, because whether a settlement be accepted or not, this system will be superseded by one involving less labour and inconvenience in the public offices." (Pedder, *op. cit.*, 426.) The entry and periodical addition of names was for the Desais of the greatest importance. The above quotation illustrates very clearly the failure of the British government officials to understand this passion for registration where, after all, the yearly allowances were minimal. Entry in the register, however, was a confirmation of, or a claim to, high origin.

It was in this respect that the Bhathelas could more easily compete with the Desais than would have been possible in an earlier period. In contrast to the decline of the latter was the economic rise of the peasant Anavils in the nineteenth century. British reports of the period portray the Bhathelas as industrious and thrifty cultivators, but their efforts were aimed above all at imitating the Desais. They tried to ally themselves to their more prominent fellow caste members by marriage. The gifts to the bridegroom's family and the wedding feast itself cost large sums of money. To organize the most sumptuous of feasts, lasting several days, was expensive but created an ineradicable remembrance and, in the eyes of the village and especially of all Anavils, served as a measure of the family's position.[31] Not only did the well-to-do Bhathelas use this device to improve their own position and create relations with the higher status group, but they also tried in general to behave like Desais. They used their earnings from the cultivation of sugar cane to enlarge their property holdings, not to increase production in order to maximize their income. A Bhathela who bought land did not do so with an eye to the market: he merely wished to be the equal of the Desais, for whom land was a prestige good rather than a capital good. A landlord's behavior determined the perspective of the Anavils in this period, and the axiom they generally lived by was: the more land, the higher the status. T. B. Naik wrote:

> It should be emphasized at this point that it is not monetary income but landed and other immovable and movable property that gives status among the Anavils. A man getting Rs 5.000/–per annum but having no land is much lower socially than another who has thirty acres of land giving an annual return of Rs 4.000/–. . . . these classes [the Bhathelas] try to get into the higher ones by getting more money and turning this money into landed property.[32]

Purchase of land alone was not enough. To conform to the life style of the Desais it was also necessary for the Anavils to limit their own

31. "Those who want to rise in social status can do so, especially through marriage into a higher family by paying a higher dowry and goods. Secondly, the classes as they rise in social ascendance spend more and more on marriage and other social ceremonies. Though conspicuous exhibitionism of one's status, *motia* or *mobhbho* (through a colorful decoration of the marriage booth, through invitations to singers and dancers, and through giving and displaying the costly gifts to the bridegroom), is the mark of all Anavil social classes, the tendency for showing off and spending increases as one rises (or wants to rise) in the social hierarchy. (The uppermost families could afford to go in for a much less elaborate affair, not requiring much money)." (T. B. Naik, 1957, 179.)

32. T. B. Naik, *ibid.*

physical work, to put an end to their wife's work in the field and to relieve her of all household duties except the preparation of meals and the care of the house altar.

This could only be done when the family took on at least one agricultural laborer in permanent service and delegated to him and the *harekwali* the most strenuous, unpleasant, and in part also impure tasks in the field and in the household. For the Bhathelas, too, it became increasingly true that esteem was accorded to him who led a comfortable life, who gave orders without having to work himself, whose wife had help in the household, who, in short, was a landlord in the same way as a Desai was. As the reports show, *halis* were recruited in times of prosperity, when a growing number of Bhathelas could afford to withdraw wholly or in part from work on the land and leave the toil to their servants.[33] When the economic tide turned, however, the number of *halis* diminished, and the masters were compelled once more to put their hand to the plow.[34] It is this alternation of extension and contraction of employment which suggests that the farm servant was to a large extent engaged for social reasons.

Although the process outlined above did much to equalize the two Anavil categories in the nineteenth century, it was chiefly the Desais who could afford a life of luxury. Their authority and wealth were undeniably affected by the government measures, but their lead was so great that only the more well-to-do Bhathelas could compete with them. Moreover, in the principality of Baroda the Desais kept their prerogatives until the end of the century. The administrative reports contain frequent complaints of their continuing influence, especially on the village level.

While the Desais clung to their privileged position, the majority of the Bhathelas was still condemned to live and work as peasants. Many of them even had to make do without the services of a *hali*. How much they were weighed down by peasant work, and its consequences, too, for their social position, is reflected in the mockery of the following Anavil song:

With an axe on his shoulder
The husband looks a Negro —

33. *Gazetteer of the Bombay Presidency*, 1877, pt. 2, 197; Shukla, 1937, 124.

34. "Halis are no longer maintained as the Anaolas find it hard enough to feed themselves, let alone their Halis." (*Papers Relating . . . Chikhli Taluka.*) The Census of 1901 states that during the famine of the preceding years many *halis* had been dismissed. (*Census of India*, 1901, pt. 18, vol. 1, 72–73.)

I wonder why
Father gave him his daughter.[35]

What was the cause of the Anavils' inclination to withdraw from all physical agricultural work? It seems reasonable to connect this attitude with their position in the ritual hierarchy as Brahmans, for whom agricultural labor, and physical work in general, are despised activities.

The Anavil Brahmans, though intelligent and brave, are looked down upon by other Brahmans. The reason in this case is that the Anavils (as they are called by other castes) are agriculturalists and their women put on shoes while carrying the men's meals to the fields.[36]

Handling the plow, especially, is looked upon as impure, and wherever in India landowning Brahmans employ farm servants, this rule is usually given in explanation. At first view a Brahman's wish to meet the demands of the Brahman ideal seems to be the chief, if not the only reason for withdrawing from agricultural work. With a farm servant at his disposal the Anavil could lead a life suitable to a member of a high caste.

As mentioned earlier in this chapter, the farm servant and his wife, both on the field and in the master's household, carried out those tasks which elsewhere are left to the members of impure castes. Speaking of functional exchangeability as an attribute of the *jajmani* system, Gould remarks:

A "clean" household cannot, by definition, retain its "clean" status unless certain kinds of work and ritual procedures with impure implications are transferred to others. If the traditional absorber of a specific form of impurity is not available, then the power and persuasion of the "clean" household must be employed to uncover someone else who will function as a substitute.[37]

In the absence of impure castes — a very low percentage of the rural population in south Gujarat—Dublas are employed to take over various less favored activities from those Anavils who have managed to improve their position.

To explain this as a desire for a Brahman way of life is only partly justified. In religious matters the Anavils are quite lax, and are not much concerned about the many rules to which as Brahmans

35. T. B. Naik, 1950, 33.
36. S. Mehta, 1930, 131.
37. Gould, 1964, 21.

they should subject themselves. In contrast to their respect for the Great Tradition as phrased in the holy scriptures they show little deference for the Brahman as village priest in their daily life. "Though Brahmans, the whole body of Anavils are laymen who never carry out priestly duties or accept alms from others. . . . The priesthood is unthinkable for them and considered below their dignity," wrote T. B. Naik.[38] Instead of aspiring to a secluded and puritanical life and recognizing obligations towards others by accepting gifts, the Anavils lord it over the village community. Their word is law for the other castes. In view of the central position they have always occupied as prominent landowners and as the local power elite, it is in fact more plausible to regard Kshatriyas as the reference group of the Anavils. In the seventeenth and eighteenth centuries they had the example of the Maratha rulers, while in the principality of Baroda the Kshatriya model was preserved in the life style of the upper stratum. As members of a dominant caste, the Anavils are comparable to rulers who have total authority over their subjects and who do not in the least detach themselves from their environment.[39]

It would, however, be going too far to explain the way of life of the Anavils one-sidedly in terms of a specific model, coupled to a specific caste which they try to imitate. The behavior of every dominant caste is unmistakably close to the Kshatriya ideal, but it does not necessarily have any reference to it. With equal justification it might be said that the ideal behavior pattern of the Anavil Brahmans was influenced by the example traditionally given by the urban elite in their territory, an elite which included not only Moslem governors representing the Moghul rulers but also members of the highest section of their own caste, the Pedivalas.[40]

If we apply this line of reasoning to the traditional village situation we may assume that the Anavil landowner's need of the services of a farm servant and his wife was in part ritually and in part socially determined. As soon as he could afford it he took a *hali* into his employ so that, as a member of the dominant caste, he could be free of physical work.

38. T. B. Naik, 1958, 389.
39. Cf. Srinivas, 1966, ch. 1, par. 3.
40. As to the style of life of the elite in preindustrial towns, Sjoberg writes, "Not only does the upper class function without engaging in physical labor, but it arrogates to itself 'luxury' items that enable it to achieve a life style that dramatically sets it apart from the lower class; in turn this ostentation reinforces its power and authority." (Sjoberg, 1965, 118–19.)

The Anavil with one or more farm servants no longer needed to work on the land and had acquired a position in which he could lead a life of conspicuous idleness, thus proving his superiority in the village community and in his own caste. *Halis* were for their master a symbol of the leisure which he enjoyed, but the master was not literally idle. He was continually busy enlarging his influence and reinforcing the commitment of his subordinates, while on the other hand he tried to damage, whether or not in coalition with others, the prestige and range of power of his peers.

In discussing the *jajmani* system I characterized the relationship between the members of the dominant caste and those of the other castes as one of patronage. The esteem enjoyed by the *jajman* was apparent from the size of his following. Landholding, chiefly concentrated in the dominant caste, was of decisive importance because, in view of the unequal distribution of property rights, it implied control of people. The clients shared in the crop, and some of them were also concerned in its production as tenants or agricultural laborers. To bind others and be assured of their dependence gratified the patron's self-esteem, but at the same time he could not dispense with it if he wished to be recognized as a prominent member of his caste and enjoy the attendant privileges.

In this light, it is not surprising that the social position of the Anavil Brahman was partly measured by the number of servants in his employ. By taking on more farm servants than he needed he reduced the proceeds of his holding. On the other hand, as his prestige was determined by the size of his property, not by the income he derived from it, he was evaluated by the number of his servants, not by the standard of whether, as a good cultivator, he knew how to use them in the most effective way in agrarian production.[41] *The question facing the Anavil landowner of the past was not how many servants he needed, but how many he could provide for.*

Extravagance in this respect was censured only when it caused the master's impoverishment. Living beyond his status, that is, taking on more farm servants than his economic capacity allowed, was

41. Writing about the production relations in Uttar Pradesh, Neale remarks, "The superior feels that he must support those who are loyal to him, whether as family head, patron, or faction leader. If he must provide for those under him, then he has no incentive to rationalize his economic operations. In effect, the followers must be paid in any case; to spread the load makes life easier and appears equitable." (Neale, 1962, 169.) In the context of this quotation, however, the author creates the erroneous impression that in such a situation there is question of a nonrational orientation.

bound to lead to loss of land, dismissal of *halis*, and in general, reduction of his group of clients. Economic decline was feared above all because it damaged the position of the landowner as patron.[42]

A massive following not only was a standard for conferring esteem upon the patron but also increased his power. This was of great importance especially in the precolonial period. Because of the unstable political situation of the seventeenth and eighteenth centuries, the revenue farmer sometimes had to extort payment by force, and only he who was able to mobilize a following if necessary could act as *desai*.

Connections with important people outside the village denoted a prominent position within it, and conversely. These Anavils, themselves influential patrons in the local community, were in their turn the clients of regional overlords to whom they owed assistance in case of conflicts. In their capacity of local power elite they profited by the weakening of central power. They began as mediators but became more and more autonomous; in the end they managed to build up a practically unassailable position and to exemplify as patrons the dominance of their caste.

In the course of nineteenth-century colonial rule, their supralocal power was curtailed. Within their village community, however, they remained completely sovereign. There, the members of the dominant caste vied with each other for the largest following in the village. In this situation of permanent rivalry, to have many followers at one's disposal was to be able to settle a feud by force if need be.[43] Under the circumstances, having several farm servants in the permanent nucleus of the client group bore witness to prudent foresight and was not at all an excessive luxury.

The power of the master rested on his ability to bind clients. It is not surprising that particularly the loyalty of the farm servants was beyond doubt. Placed low in the social and ritual hierarchy, little specialized and held in low esteem in their work, selected from an abundant supply, continually employed on the land or in the house-

42. A statement by Lorenzo concerning the *kamias* in Chota Nagpur, comparable to the *halis* in south Gujarat, also points in this direction: "The sale of the *Kamia* is supposed to be derogatory. It is often the result of the landlord's uncertain economic position, which makes it difficult for him to maintain his Kamias. The Babhan mahajans [the caste of the masters] take much pride in having a greater number of Kamias, and often as a result of this race for Kamia ownership, their sale is postponed as long as possible." (Lorenzo, 1943, 80–81.)

43. See also Cohn, 1955, 65.

hold of the master and therefore always at his disposal, these people were bound hand and foot.

In contrast to the members of the artisan and serving castes the Dubla servants had only one patron, of whom they were both the least favored and, perforce, the most loyal clients. The least favored since they carried out strenuous physical work and since, in a community in which obligations to others are to be avoided, their attachment to a master implied an extent of dependence that was looked upon as very demeaning.[44] If in chapter 3 the members of the dominant caste were called the patrons *par excellence*, the dependence and subordination of clients is preeminently represented in the category of farm servants. In the society of the time, no greater distance existed than that between the dominant Anavils and the lowly Dublas. The comprehensive inferiority of the latter lent even more luster to the independence and superiority of the Anavil masters.

A patron-like behavior was especially embodied in the way of life of the Desais. The system of prestige and power, in which the patronship of the members of the dominant caste was expressed, largely functioned in relation to the village community.[45] Because, by the end of the last century, landholding within the dominant caste became more equalized, an increasing number of Bhathelas successfully competed with the Desais. "Lordly behavior" and a "kingly style," or whatever such an orientation is called, has traditionally been the behavior that set the tone and was a worthy aim of Anavil Brahmans. In this century, however, it has become attainable by a growing number of the Desais' fellow caste members.

Administrative reports of the nineteenth century described the Anavils as rulers over the village population. Again and again we find complaints of the pressure exerted by them, and they are referred to as impertinent and quarrelsome. What the administrators regarded as stirring up trouble against the British and simple griev-

44. "Occupations involving heavy physical labor are generally esteemed to be very low and direct dependence on others for one's livelihood suggests lack of freedom, which is also not a sign for a high status occupation in the village." (Joshi, 1966, 28.)

45. Cohn represents the village as a small principality with a closed prestige and power system in the traditional community. "Before the end of the nineteenth century, the little kingdom was an almost closed prestige system; prestige depended on the amount of land one inherited from his ancestors, the status of his family in relation to the founding ancestors, and the low caste followers that he could muster." (Cohn, 1959–60, 92.) It is a characterization of local dominance in north India which also applied to the Anavils of south Gujarat.

ance-airing was really a symptom of the dominant position of the Anavils and of their mutual rivalry. Though intolerant of criticism directed against themselves, they are extremely outspoken and emphatic in their pronouncements on any subject, whether it concerns them or not. Their behavior is in marked contrast with that of lower castes, which as a rule is characterized by humble wariness and the fear of unnecessarily committing oneself and irritating one's superiors. The bossiness of the patrons, reflecting their arrogance and self-importance, contrasts sharply with the sometimes abject servility of the clients. In the presence of British officials, the patrons in their turn had to feign submissiveness, which evidently they did not find easy.[46] In the village system, however, the Anavils are accustomed to command, not to obey, and when they speak, the other villagers are expected to listen submissively. The Anavil's behavior displays irascibility and superiority. He is wont to dwell emphatically on the stupidity and backwardness of other castes, whether members of those castes are present or not. It is a consequence of their behavior

46. A small anthology of statements concerning the Anavils: "They are notoriously a quarrelsome and intriguing race, full of duplicity and chicanery, adepts in controversy, most obstinate in purpose, and extremely fond of litigation. They will dispute an usurped right, or the doubtful possession of a field or tree with a pertinacity unequalled even among the natives of India. Neither reason nor argument can convince them, and after petitioning every tribunal of appeal, I believe death alone puts an end to many of their inveterate disputes about lands." (Bellasis, 1854, 2.)

"I would have passed over any allusion to the character of the Bhatellas, were they not a power in the community, exercising great influence over the minds of the people who are numerically stronger than themselves." Captain Newport thus portrays them in 1822: "My intercourse with the Bhatella Brahmins has been extensive and intimate, and as individuals I found them obsequious and servile when they expected benefit; would seldom or never answer a question direct until its object had been guessed at, and then guided by interested motives, would either evade it by pretended ignorance or by direct falsehood. Indeed experience has proved to Government that some of the most respectable are not trustworthy."

"It rejoices me to think that both you [the Collector of Surat] and the revenue commissioner are not unacquainted with the turbulent and never-to-be-satisfied class of *ryots* in this district. Villagers who had not petitioned were threatened with excommunication, and in this manner it has been the Bhatella policy from the commencement of our rule to show as much general dissatisfaction as is possible, in the hope that per adventure their efforts may bring forth a little good." (*Papers Relating . . . Jalalpor Taluka*, appendix R [The first survey and settlement report, of 1868–69], p. 30.)

"These Bhatelas are notorious as the greatest intriguers in the country. This character, assisted by the hereditary influence I have described them in [sic] possessing, has enabled them to worm themselves into positions of power as patels of villages, which they have used to the injury of the revenues by exacting labour from the lower castes of cultivators, and thus preventing them to improve their own condition." (*Papers Relating . . . Chikhli Taluka*, 63.) Where the authors of these reports speak of Bhathelas it appears from the context that by this term they refer to all Anavils.

on the local level as *Rajas*, a term which they still acknowledge with a condescending smile.

4.5 The Dubla Farm Servant as Client

Earlier in this chapter it was pointed out that the *hali* could claim total though extremely sparing support from his master. As a *kamin* received his grain allowance irrespective of the service he had performed, so the *hali* was not remunerated on the basis of his merits but on that of his needs, which were rated low indeed. There being no contractual agreement related to a market situation, the allowance of the farm servant was not very flexible. This allowance—a more accurate term than "wage"—was paid largely in kind and remained at the same minimal level throughout the last century. The patronship of the master was founded on the bondage of the *hali*. What he paid his Dubla farm servants always exceeded what they set against it in the form of labor, at any rate in his own calculation. The debt element, however, has been given undue importance in the literature. Writing about a similar service relationship in Uttar Pradesh, Singh remarks,

> The debt of a *harwaha* [comparable to a *hali*] is a fictitious asset of the money-lender. These people never take their debt seriously, for they know that their incomes being grossly inadequate redemption of debt is out of the question. They look upon the money that they are able to borrow from the Zamindar as their "income." And the Zamindar too does not consider the money advanced to a *harwaha* a loan. He does not want the debt to be redeemed.[47]

As this quotation shows, instead of wishing to terminate the relation as soon as they could, both parties aimed at continuing it as long as possible. As already mentioned, the master did not want the debt repaid. He did not make a "loan" to press for payment afterwards, but rather to be able subsequently to assert himself as a patron. He made it very clear who was the giving and who the receiving party, and in this way impressed both gratitude and loyalty upon the farm servant. The size of a servant's debt was a measure of his commitment, of the control which could be exerted on him by the Anavil. The master repeatedly, and often on the basis of false figures, told the Dubla that he got much more than he was entitled to. In this

47. Singh, 1947, 134.

way he managed at once to demonstrate his liberality and to emphasize his servant's dependence, in short, to behave like a patron.

Nor did the *hali* work to pay off his debt, as has often been suggested, for to be an unattached agricultural laborer was the last thing he wanted. Instead of keeping his debt low, the farm servant tried to exploit his privileged position and move his master to do him as many favors as possible. He appealed constantly to the Anavil's beneficence. By working on his sense of honor as a patron, by badgering him for extra compensation and pressing for all kinds of remuneration, he stimulated the master, as it were, to be liberal and to make a show of behaving like a good master. The farm servant was undoubtedly aware of the fact that by so acting he increased his dependence, but he accepted it as inevitable and considered himself entitled to protection.[48]

On the other hand, providing for the material needs of his farm servant was not the only obligation of the master. The protection offered by the Anavil to his *hali* reached far beyond giving him a living: he had to guarantee his existence in a much wider sense. This meant that the master should defend the social interests of his servant and that he was held responsible to some extent for all the actions of his subordinate.[49] Consequently, the *hali* of a prominent Anavil patron was assured of a reasonable amount of security and protection. For his part, he had to behave loyally, that is, do nothing that might provoke the displeasure of his master. Should conflicts arise, he was expected to take his master's side, even if it meant opposing his fellow caste members, the Dubla clients of the other Anavils.

As has been explained, the client enhanced the prominence of his patron. Apart from loyalty, the Dubla owed his Anavil master respect and submissiveness. His humbleness in word and gesture alone were sufficient to accentuate the superiority of the master. Old Anavil informants spoke to me of the great deference and timidity with which they used to be approached by the Dublas, while they censured severely the "impertinence" and "unmannerliness" of the agricultural laborers of today.[50]

48. "From close enquiries I have made, I can say with certainty that these poor people [*halis*] are treated with kindness and consideration, and are looked upon as humble dependents who have great claims on their masters and their families" is a comment from *Papers Relating . . . Chikhli Taluka*, 51. See also *Gazetteer of the Bombay Presidency*, 1883, pt. 7, 118.

49. See also Srinivas, 1955, 28.

50. Cf. Cohn, 1955, 61–62.

The Anavil's superiority was reinforced in a religious context when, for instance, at a marriage or a death in the house of his patron, the *hali* participated in the ceremonies in a clearly subordinate and servile way. This behavior was both a solemn confirmation of the tie between master and servant and an indication of the distance which separated them.

The dependence of the Dubla servants far exceeded that of the patron's clients who belonged to other castes. The Dublas had to rely on the benevolence of their masters for everything. The farm servant had no one else to turn to for favors; his Anavil master was the only one who was prepared to help him if he was in difficulties, but at the same time no one else could thwart and harass him as the master could.[51] Naturally this far-reaching bondage in the purely economic sense helped to inculcate loyalty and discipline in the farm servant, but it would be incorrect to suppose that only pressure could have induced him to go to such lengths.

Considering the weak position in which the *hali* found himself in all respects, it was in his own interest to side with his Anavil patron. By leaving no doubt of his loyalty, at least in his behavior, the Dubla hoped in his turn to compel his master to protect him. If in his attitude and behavior the farm servant also honored his Anavil as a benefactor in an exaggerated way, he did it partly to remind him of his obligations as a patron and to spur him to magnanimity.

Moreover, and this is characteristic of a relationship of patronage, the Dubla strengthened his own position in relation to third parties by identifying himself emphatically with his Anavil. The esteem and influence enjoyed by the master was shared by the servant, for a little of the farmer's importance was reflected upon his subordinate.

Earlier in this chapter it was maintained that bondage was attractive to the Dublas on account of the economic security it offered. Because of their contacts with the rich and powerful villagers and the privileges and protection that these contacts afforded them, the Dubla farm servants also rated higher in social esteem than their fellow caste members. The unattached laborers among the Dublas, some sources state, were even despised.

51. "Fear of eviction from the house site, fear of recall of loans, fear of loss of the additional income from the piece of land gifted to him and lastly, the security taken from a farm servant in the form of deferred wages to be paid only at the harvest, are more powerful and terrible sanctions than what the ordinary law may provide." (Sivaswamy, 1948, 30.) See also Beidelman, 1959, 36–37.

It has become a matter of prestige in the Dubla community to be a Hali, to work for a Dhaniamo. As a woman without having married a husband has no prestige in society, similarly a Halpati without a Dhaniamo as his master has no prestige in his community.[52]

The loyalty of the farm servant was said to be a byword. In the stereotype so stubbornly maintained by Anavils, the Dubla would go through fire and water for his master and identified his master's interest with his own. He could be counted upon to carry out as well as he could any instruction the master chose to give him. Attachment to a prominent Anavil gave the Dubla security and prestige, and the Dubla was devoted to him, if only for this reason. According to some authors, he was filled with contempt for those who deserted their *dhaniamo*.

Loyalty to the master figured as an important item in their moral code. They looked down upon those who were not and very often would not have anything to do with such deserters. The master on the other hand always appreciated this loyalty and put a premium on it by raising the status of his faithful dubla. There was not much of material reward but the labourer was admitted more and more into the confidence of the master, given a freer hand in his work, allowed a sort of familiarity and intimacy with the master's family and was trusted with work of a more delicate nature than that of a purely agricultural labourer.[53]

Owing to the frequent and intensive contacts between master and servant their relation lost much of its rigidity and stiffness, and came to bear a markedly personal stamp. The *hali* was the master's confidant. Being charged with the routine work he bore more responsibility than the casual laborers. Regarding all important agricultural questions the servant was consulted by his master, and the trust the latter put in him was apparent from his independence in carrying out orders. The *hali* knew about all kind of intimate family details of his Anavil which were hidden from outsiders and had to remain so. Even now every Anavil family can illustrate with incidents from the past the affection and loyalty of its former Dubla servants, which incidentally also enhance the glory of its own patronship. Such moving stories about the geniality governing the service relationship can usually be traced to an Anavil source and are not very convincing, if only for this reason.

52. J. Dave, translation from the Gujarati, 1946, 18. For a remark of similar purport, see *Report of the Hali Labour Enquiry Committee*, 1950, 48.
53. Kishore, 1924, 427.

4.6 Halipratha: *A System of Unfree Labor Mitigated by Patronage*

It is not difficult to recognize that servitude cannot have been so harmonious a relationship as is sometimes suggested in retrospect. The stress was on the "good" master and the "good" servant, that is, on the Anavil who carried out his obligations without stint or chicanery, and the Dubla who served his master loyally. Rather than the actual behavior it was the ideal relationship of *dhaniamo* and *hali* that was described, presumably on the strength of Anavil sources. Obviously the Anavils wished to create the impression that they treated their servants excellently. As patrons, moreover, they wanted to leave no doubt of their clients' loyalty.

Nor did the Dublas themselves, on being interviewed, give any evidence of dissatisfaction, sometimes to the rather naive-sounding astonishment of British government officials. As I explained, bondage was accompanied by many advantages, which partly compensated for the lack of liberty. But the servant had to avoid any semblance of impertinent action and behave in all respects as a subordinate and dependent. Open doubt of the master's good intentions, denial of his pretensions, was highly improper because it detracted from the Anavil's dominance. The *hali* should show himself satisfied and obedient. However, if the master refused to meet the obligations which had been established by custom, the *hali* could remind him of them by a show of indifference and indolence, or by remonstrating with his master in front of others. On the other hand, if the Dubla exceeded the limits of self-defense it might cost him dear.

Many reports of the nineteenth century asserted that the servants rarely ran away, but this, too, should not be adduced as proof of cordial relations between them and their master. To have a master meant for the Dubla to belong to the privileged category among the Dublas. The *hali* who deserted his master relinquished his social security, which he would do only in extremity. And where else could he have found work? In the traditional society, employment outside agriculture was extremely scarce. Moreover, the local isolation, a consequence of the poorly developed system of communications, made it difficult for him to judge the possible demand for labor elsewhere. But if alternative employment suddenly materialized nearby, he seized his chance with alacrity. In the eighteen-sixties, for instance, many farm servants left their masters when labor was needed for the construction of the railroad through the plain of south Gujarat.[54]

Such unexpected and temporary possibilities of escape aside, a *hali* decamped only if feelings had run too high. Practically the only way in which he could defend himself was by trying to withdraw completely from his master's authority. Precisely because this meant a total breach it was a step that the farm servant certainly would not take lightly, and moreover, it little profited the Dubla. At best he might succeed in finding a better master, but this prospect was too uncertain to warrant his severing the relation hastily.

Sometimes, however, the *hali* ran away on the instigation of another Anavil. As already mentioned, the relation between the most prominent members of the dominant caste as patrons was one of mutual rivalry. The master who could not keep his servants lost prestige. This was the more painful for him if a *hali* had been induced to leave his master to serve another Anavil in his own or a neighboring village. The code of the Anavils forbade such illegal transfer of a servant, and their caste *panch* saw to it that the rule was observed. The Anavil who engaged a *hali* who had left another master without the latter's permission was fined, and the Dubla himself was punished by caning. Yet it is difficult in retrospect to judge the effectiveness of these sanctions.

When a breach was not due to the struggle for power between competing patrons with the *hali* as prize, the *hali* could do little but fly from the village. Usually he went to the village of his mother or his wife, but he was not safe there. When informal collaboration between the Anavil Brahmans did not lead to his being brought back, the local authorities helped to trace runaway *halis*. In the first half of the nineteenth century, running away even rendered a *hali* liable to punishment by law.[55] Later, complaints about disloyal Dublas were no longer admitted, and officially the master could not do much against a runaway servant. The actual situation, however, had not changed a great deal. The Anavils remained the most important power elite in the principality of Baroda, and in the British

54. Enthoven, 1920, 347.

55. J. Vibart, Principal Collector of Surat, in a letter dated 16–12–1835, wrote, "By a letter from Government, dated 19th April 1822, the Magistrates are authorized to apprehend and return to his master any Halee who may abscond." G. Grant, Acting Judge and Session Judge in Surat, expressed the same view in a letter dated Feb. 22, 1836. This correspondence was included as Appendix to Report on Slavery in *Account and Papers*, vol. 16, session 15-11-1837 — 16–8–1838, pt. LI; see fn. 1 to this chapter.

part of south Gujarat, police functionaries and other local government servants were recruited from their midst. This meant that the masters could still informally appeal to the local authorities to compel a runaway Dubla to return.

Thus the Anavil landlords seem to have had the *halis* completely in their power. The members of the tribal and landless Dubla caste had no choice but to put themselves in the keeping of a master. This explains at the same time why the relationship began and continued with the approval of the servant. The tie attaching him to the master became increasingly close, and only on the master's initiative could it be severed. It is true that in the course of time the amount of support to which the *hali* was entitled had become standardized, but if necessary he had to accept less. The exceedingly unequal division of mutual rights and obligations was due to the difference in position between *dhaniamo* and *hali*. In theory, the right of the servant was the obligation of the master, but the former was not guaranteed in any way. The great economic, political, and social power of the Anavil made it possible for him to dictate, and in any case interpret, the service conditions one-sidedly. His obligations were rather in the nature of favors to be granted as he thought fit. The paternalistic attitude of the master could barely conceal the elements of enforcement that were inherent in the relationship. The servant was at his master's beck and call. Together with his family he had to comply with the latter's every wish, which might include sexual intercourse with the *hali*'s wife.

On the other hand, the element of patronage prevented any pronounced tyranny and complete exploitation, which would have detracted from the esteem in which the Anavil was held. Then, too, he had to rely on the support of his following. The patron's position stood or fell by his willingness to provide for and aid a group of followers, to let them participate in his goods, his power, and his esteem. The master must carry out a minimum of the obligations he owed to his servant. In the end, it was on the latter's labor that the success of the crop depended. In the traditional agrarian system, patronage not only served as a corrective of the lopsided distribution of scarce goods in traditional village India, but also accentuated inequality in all spheres of life. The presence and the symbols of subordination of the clients enabled the patron to demonstrate his superiority. As the economic and ritual distance between them widened, the relationship came to contain an element of exploitation, which in the case of the low-classed agricultural laborers was mitigated only by the mutual intimacy and sympathy inherent in the personal, usually hereditary tie with the landlord's family. In the concept of *patrimoniale Herrschaft* (patrimonial rule), Weber, perhaps harking back to his investigation into the conditions of agricultural laborers in the German provinces east of the Elbe River at the end of the last century, characterized the relationship in the following words:

The master "owes" the subject something as well, not juridically, but morally. Above all — if only in his own interest — he must protect him against the outside world, and help him in need. He also must treat him "humanely," and especially he must restrict the exploitation of his performance to what is "customary." On the ground of a domination whose aim is not material enrichment but the fulfilment of the master's own needs, he can do so without prejudicing his own interest because, as his needs cannot expand qualitatively and, on principle, unlimitedly, his demands differ only quantitatively from those of his subjects. And such restriction is positively useful to the master, as not only the security of his domination, but also its results greatly depend on the disposition and mood of the subordinates. The subordinate morally owes the master assistance by all the means available to him.[56]

In south Gujarat it was the Anavil Brahmans who, as the dominant landowners, regulated the distribution of agrarian produce. It would be wrong to conclude that instead of production based on profit for a few, fulfilment of the material needs of all was aimed at. The Anavils were not altruistic to that extent. A more plausible explanation is rather that, for the landlords, maximalization of income took second place to maximalization of prestige and power, which in the nineteenth century amounted to behavior as a patron in the village. This objective required a minimal fulfilment of the need of clients for aid and protection. At that time, the Dubla *halis* were objects in the endeavor of the members of the dominant caste to acquire more power and esteem, not subjects in a surplus-oriented

56. Weber, 1962, 682. Concerning the past situation of agricultural laborers in the eastern regions of Germany, Weber remarks in another publication: "Except in times of extreme political turmoil, a class consciousness among the rural proletariat directed against the masters could only develop purely individually with relation to one master alone insofar as he failed to show the average combination of naive [sic] brutishness and personal kindness. That, on the other hand, the agricultural laborers were not normally exposed to the pressure of a purely commercial exploitation was in complete accordance with this. For the man opposite them was not an 'entrepreneur' but a territorial lord in miniature." (Weber, 1924, 474.)

market economy. An important factor in this process was the effort of the Bhathelas to attain the same position as the Desais partly by taking *halis* into their employ, which enabled them to imitate the life style of the higher status group.

On the basis of all these considerations it can be concluded that servitude in south Gujarat during the nineteenth century was essentially a form of unfree labor that was complicated and mitigated by a relationship of patronage.

Chapter 5

The Disintegration of the *Hali* System

5.1 Changes in the Relationships between Landlords and Agricultural Laborers

During the past few decades, drastic changes have taken place in the relations between landlords and agricultural laborers in south Gujarat—changes which, in fact, amount to the distintegration of the *hali* system. We read in numerous reports that Dublas offered themselves as farm servants with growing reluctance or, if they were employed as such, that they ran away from their masters. For their part, the Anavils began to be less inclined to take on farm servants, and came to prefer casual labor, which they could engage and dismiss according to their needs. Similar developments are reported for other regions in India, and here, too, the greater freedom of the agricultural laborers is stressed and contrasted with their attachment in an earlier period.[1]

The decline of the *hali* system was gradual. There was certainly no sudden turn in the relationship between Anavils and Dublas. Nor is it readily conceivable that such an important institution in the social and economic life of a large part of the agrarian population should come to an end within a short span of time. Although complaints about runaway *halis* have never been scarce, it was from the beginning of the twentieth century onwards that mention was made of increasing friction between masters and servants. Time and again we read in such later reports that servitude has fallen into disuse, or at any rate will very soon be a thing of the past.[2] Nevertheless, in publications from the nineteen-thirties and forties, the Dublas were still called unfree laborers, in the employ of the large landowners.

1. See, for instance, the report of the eighth conference of agricultural economists, which was devoted to systems of wage payment in agriculture. Published in *I.J.A.E.*, III, 1948, no. 1. Cf. also Thorner, 1962, ch. 3.

2. *Census of India*, 1921, vol. III, pt. 1, 222; *Gazetteer of the Baroda State*, 1923, vol. I, 352; *Census of India*, 1931, vol. XIX, pt. 1, 254–55.

Occasionally these accounts contain the prediction that there was no reason to expect that the situation will soon change.[3] It appears that the latter view was the correct one, as was evidenced by the fact that in 1947, after India had become independent, the government of the then State of Bombay appointed a committee to inquire into the situation of the *halis* in south Gujarat.[4]

If anything becomes clear from all this source material it is that the change observed in the relationship between Anavils and Dublas was a gradual one. It is therefore impossible to say precisely when there was no longer question of a *hali* system. There are indications, however, that Anavil-Dubla relations changed most drastically after the First World War. According to Joshi, the *hali* system disintegrated between 1930 and 1950 in his village of field work. In another village situation, I. P. Desai believes this change took place between 1940 and 1960.[5] Their conclusion could be considered to apply to the whole of south Gujarat.

Although in the literature we find many more or less elaborate discussions as to how the disappearance of the *hali* system should be explained, it is striking that the authors dealt very briefly with the labor relationship that replaced it. Most of them confined themselves to the statement that day-laborers came to form an increasing percentage of the total labor force. As was pointed out earlier, that category was never absent in the traditional society, but the point is that their proportion rose sharply in recent decades — a development which also took place elsewhere in India. The authors refer only incidentally to the recruitment of these casual laborers and their labor conditions, and more generally to the nature of their contacts with employers. They create the impression that servitude was replaced by a casual and intermittent service relationship which was entered into only for limited periods varying from one day to a season. In this view, the close and permanent personal tie that had existed between employer and employee was replaced by a superficial and ephemeral contractual relationship.[6]

In part III of this study will be a found a detailed description, based on materials gathered during field work, of the pattern of re-

3. Shukla, 1937, 132–3; *Census of India*, 1941, vol. XVII, 74.

4. *Report of the Hali Labour Enquiry Committee*, delivered by M. L. Dantwala and M. B. Desai. Not published.

5. I. P. Desai, 1964, 162.

6. "During the fourth and fifth decades of this century, the method of permanent 'employment' of Halis gave way to casual labourers on a seasonal wage basis. This made the village labour market more or less 'free' " (Joshi, 1966, 57.)

lationships that has come to prevail between Anavils and Dublas in south Gujarat. At this point I intend only to sketch the circumstances under which the *hali* system disappeared. After a discussion of the factors which, in the literature, are regarded as responsible for this disappearance I shall briefly outline my own interpretation of the change that has taken place in the relationship between landlords and agricultural laborers.

5.2 Current Views on the Disappearance of the Hali *System*

To explain the disappearance of a system of unfree labor — and the *hali* system has always been described in these terms — the obvious theory to apply is that of Nieboer, which was commented upon in chapter 2. His thesis can be summarized in the statement that an abundant supply of laborers obviates the necessity of their bondage. In such circumstances the landowners no longer need to protect themselves against shortage by assuring the permanent bondage of one or more agricultural laborers. This explanation of the disappearance of the *hali* system conforms with the widespread assumption that, owing to a process of agrarian pauperization under colonial rule, the percentage of agricultural laborers has risen sharply. As I said before, this assumption needs some modification. The returns of the successive censuses — its chief foundation — are too unreliable to be compared without correction, one of the reasons

TABLE 5.2.1

SHIFTS IN THE NUMBER OF AGRICULTURAL LABORERS, 1881–1921

Year of census	*Agricultural laborers*	*Remaining agrarian population*	*— (percent of total agrarian population) Agricultural laborers*
1881	259,205	—	—
1891	190,896	1,196,580	14
1901	372,964	634,688	37
1911	313,479	970,675	24
1921	295,815	1,058,182	22

being a varying definition of the concept of agricultural laborer. The same is true of the data concerning south Gujarat, as is evident, for example, from the figures for the state of Baroda.[7] The size of the

7. Data taken from *Census of the Baroda Territories,* 1881, and *Census of India,* 1921, vol. XVII, pt. 1, 364.

category of agricultural laborers, as indicated in the table, shows improbable fluctuations. Depending on the point of view chosen by the interpreter, their percentage of the total agrarian population can be said to have either fallen or risen in the course of time. A number of socioeconomic village monographs from the first half of this century contain more detailed information to go by.[8] However, although it can be concluded from them that land was increasingly fragmented, there is no evidence to show that in the end landholding groups became landless on any great scale. There are no indications of important shifts in land ownership during that period. A few smaller castes aside, Anavil Brahmans remained the most important landowners.[9] Their economic dominance in the plain of south Gujarat continued to be uncontested. Nor did there occur any drastic change in the position of the Kolis, most of whom continued to be peasants and tenants. It is true that they occasionally worked for the larger landowners, but such an additional source of income for them had already been noted in mid-nineteenth century reports. As before, it was especially the Dublas who worked as agricultural laborers. In short, there is no reason to assume that landowners or their descendants have become landless in any considerable number since the beginning of this century in south Gujarat.[10]

Growth of the proportion of agricultural laborers could, however, also be due to a sharp increase of the social category from which they were traditionally recruited. To test this assumption I have compared, in the following table, the rise of the number of Dublas in the state of Baroda with the population increase of their most important employers, the Anavil Brahmans.[11]

The two population categories show a clearly different demographic pattern. The numerical strength of the locally dominant caste has changed very little in the course of five decades. Birth control to lessen the risk of many daughters,[12] the prohibition for

8. Especially Mukhtyar, 1930, and Shukla, 1937.

9. Joshi (1966, 67) calculated that in 1901 the Anavils in his village of field work possessed 90.4 percent of the land, and in 1957, 93.1 percent.

10. Some reports state that, at the beginning of this century, landless laborers became owners of the unreclaimed, usually infertile village land. (*Land Revenue Administration Report*, 1923–24, 37; *Census of India*, 1921, vol. XVII, pt. 1, 360–367.)

11. The data have been taken from the 1881 *Census of the Baroda Territories*, 1883, 143; *Census of India*, 1921, vol. XVII, pt. 1, 344; *Census of India*, 1931, vol. XIX, pt. 1, 431.

12. *Census of India*, 1901, vol. XVIII, pt. 1, 234; see also *Census of India*, 1911, vol. XVI, pt. 1, 278, and Mukhtyar, *op. cit.*, 56.

TABLE 5.2.2
NUMERICAL INCREASE OF DUBLAS AND ANAVILS IN THE STATE OF BARODA, 1881–1931

Year of census	*Dublas*	*Anavils*	*Ratio of Dublas to Anavils*
1881	42,197	10,335	4.0
1891	48,886	11,148	4.4
1901	41,043	10,862	3.8
1911	50,623	9,916	5.1
1921	51,834	10,751	4.8
1931	65,459	11,818	5.9
1941	76,479	—	—

widows of this high caste to remarry, and, especially, migration are the chief factors that explain the stagnation of the Anavils in their native region. The landless Dublas, on the other hand, increased numerically by leaps and bounds. Although it was they and other economically weak castes who were most affected by crises, such as the famine at the turn of the century and the influenza epidemic of 1917,[13] the relatively higher birth rate of the lower castes combined with a gradual long-term decrease of mortality led in the end to an accelerated growth of this least-favored group. As the census returns of 1921 showed, the number of children per 1,000 married women between fifteen and forty years of age was much higher in the tribal castes than in the castes of Brahmans, being respectively 240 and 188.[14]

It could be concluded from this outline that, in consequence of a demographic shift, the supply of agricultural laborers gradually increased to such an extent that the Anavil landlords no longer needed to bind Dublas as farm servants (see table 5.2.2). Did the *hali* system, in other words, therefore disappear as a result of the accelerated expansion of the population of agricultural laborers?

A first objection to this line of argument is that the rise of a surplus in the first decades of the twentieth century would imply a

13. *Report of the Famine Operations in the Baroda State for the Year 1899–1900*, 1900, 3; *Census of India*, 1901, vol. XVIII, pt. 1, 231; *Census of India*, 1921, vol. XVII, pt. 1, 37, 42.

14. *Census of India*, 1921, vol. XVII, pt. 1, 212; also *Census of India*, 1931, vol. XIX, pt. 1, 115. To my knowledge, the demography of the caste system has not been much researched, although the social consequences of unequal numerical increase have great weight.

previous shortage of agricultural laborers. Now this I refuted emphatically in the foregoing chapter. At most we can speak of periodic scarcity of labor, but in this respect the situation did not change. Complaints on the part of landowners that there were not enough hands at the peak of the season to cope with all the work that had to be done were a recurrent theme.[15] If it was true that, in the traditional society, the *hali* system did not function in circumstances in which there was a shortage of agricultural laborers, it may be too hazardous to explain its disappearance as being the result of a surplus.

To summarize, there are not enough indications of a process of agrarian impoverishment leading to a rise in the percentage of landless laborers. We do know that, owing to a differential demographic pattern, the proportion of agricultural laborers within the agrarian population has increased, while on the other hand the dominant caste has lagged far behind in population growth. It is undeniable that this change in numerical ratios has affected the relations between Anavils and Dublas, but in my opinion the suggestion of a direct connection between the growth of the category from which the agricultural laborers were mainly recruited and the disappearance of the *hali* system does not take into account all the facts.

It is therefore not surprising that the Anavils' tendency to stop employing attached laborers is usually attributed to completely different reasons. Of these, the cost argument is the most important. Because of a sharp rise in wages—they are reported to have doubled within a span of ten years, from 3.6 annas in 1907 to 7 annas in 1917—few landlords would have been able to provide for one or more laborers throughout the year.[16] The amount to be paid by the master when the *hali* entered his service also rose considerably, while owing to the pull of urban centers the risk of the *hali* running away was much greater than before. Finally it is suggested that subdivision of the land had affected the landlord's financial capacity.[17] Especially in the crisis years after 1930, many of them were obliged to dismiss their farm servants.

Yet this line of argument, which presupposes a deterioration in the landlords' economic standing, is equally unconvincing. In the

15. Mukhtyar, *op. cit.*, 159–61; M. B. Desai and C. H. Shah, 1951, 49.
16. *Census of India*, 1921, vol. XVII, pt. 1, 42; see also *Census of India*, 1931, vol. XIX, pt. 1, 255.
17. *Papers Relating . . . Jalalpor Taluka*, 1900, 8. This argument is difficult to reconcile with the slight numerical growth of the largest group of landlords, the Anavil Brahmans.

first place, the statements about the large wage increases should not be taken too seriously. It was the money value of the allowance that increased, for the allowance itself continued to be paid largely in kind for a long time and certainly did not increase.[18] This was only slightly less true of the "loan" contracted by the farm servant at the beginning of his service. This loan was usually meant to cover the expenses of his marriage, and as a rule the master himself continued to supply the necessary food and other requirements for this purpose.[19]

As production for the market expanded — this was especially the case after the First World War — the proportion of the total wage that was paid in money did increase a little. But owing to the rise of prices of articles which the farm servant formerly was given but now had to buy — salt, tobacco, cooking fuel, etc. — this partial changeover to a money wage was not detrimental to the employer. Again and again we read that the rise in the wages of agricultural laborers lagged behind that of the cost of living.[20] The same is true of the allowance received by the servant on the occasion of his marriage.[21]

The economic crisis of the nineteen-thirties may have accelerated the disintegration of the *hali* system. Whereas formerly *halis* are said to have been dismissed in years of adversity but taken on again when times were better, the farm servants were now no longer taken on at all. Nevertheless, the alleged weakening of the economic position of the landlords does not, in my view, offer a satisfactory explanation of the fact that bondage fell into disuse.

Much more plausible is the finding that, because of a drastic change in the crop system, the landlords no longer needed to use permanent labor. In a village study of 1927, Mukhtyar posited a connection between the cultivation of sugar cane and the existing system of labor relations.

In this village, and, in point of fact, in the whole of Southern Gujarat, there still obtains a system of labour called the Hali system in which a

18. J. M. Mehta, 1930, 127; Shukla, *op. cit.*, 123–28.

19. "The loans in kind form a substantial proportion of the total advance, as almost the entire quantity of foodgrains, pulses and vegetables required for feasting on the occasion are advanced by the Dhaniamo. The other significant factor about the marriage loans is that as compared with the rise in the prices and the cost of living, the increase in the total computed expenses on the Hali's marriage has been much lower." *Report of the Hali Labour Enquiry Committee*, 26.

20. *Census of India*, 1931, vol. XIX, pt. 1, 269; *Report of the Hali Labour Enquiry Committee*, 30; C. H. Shah, 1952, 440–61; Shirras, 1922.

21. See fn. 18 above.

labourer mortgages his labour to the farmer for a loan he takes for celebrating his marriage. A capitalistic cultivator keeps one or two Halis for performing field operations. He is bound to maintain them whether he exacts work from them or not. He, therefore, deems it wise and profitable to occupy them in sugarcane-cultivation.[22]

The decline in the cultivation of this highly labor-intensive crop after the First World War led to vastly diminished demand for permanent labor. Joshi, in the history of another south Gujarat village, also sketches a contraction of agricultural activities after the nineteen-thirties, when Anavils first began to cultivate mango trees. Whereas laborers formerly were needed throughout the year for sugar cane and garden crops — such as ginger, spices, root crops, and bananas — most of the work could now be performed by temporary hired laborers in the picking season. The new crop was introduced only gradually, and this prevented a sudden end to bondage. New *halis* were no longer engaged. Dublas who were employed as farm servants remained in service.[23] On the basis of these data it is plausible to say that the cause of the disappearance of the traditional service relation should not primarily be sought in the impoverishment of the Anavils, but rather in the changeover to a crop system not entailing very intensive labor.

However, it would be incorrect to create the impression that the landlords took upon themselves the initiative of abolishing the *hali* system. A number of reports hold the opposite view, that is, that the change in the labor pattern was forced upon them. In the preceding chapter I explained that bondage was certainly not one-sidedly imposed on the Dublas. On the contrary, by becoming *halis* they were assured of at least a minimal existence in the more or less closed village economy. With increasing opportunities for employment outside the village, servitude lost much of its attraction for these landless laborers. Fewer and fewer Dublas were prepared to enter into an agreement which, in practice, bound them for life.

As time went on, the possibilities of finding a livelihood outside agriculture for themselves in the region gradually increased. After a period of stagnation in the last century, the cities of Surat and Navsari began to prosper again, and smaller towns along the railroad began to develop into industrial centers of some importance.[24] Moreover, seasonal migration over longer distances contributed

22. Mukhtyar, *op. cit.*, 97.
23. Joshi, *op. cit.*, 57.
24. See, for instance, Kapadia, 1966.

considerably to the expansion in employment. On a small scale at first but in ever-increasing numbers, especially since the Second World War, agricultural laborers and small peasants migrated during part of the year from the south Gujarat villages to the environs of Bombay. They worked there in the dry season—the period between November and the end of May—in salt-pans and brickyards in the open air. It might therefore be concluded that the Dublas felt less and less attracted to agricultural servitude. The same line of thought led to the observation that they no longer relied on the support of the landlords who enabled them to marry; it should be borne in mind that precisely the essence of the *hali* system has been deemed to be the necessity for the *hali* to take a "loan" for that purpose. The Dublas began to feel that complete dependence on the master, which they formerly accepted as security, was a heavy burden. This is the tenor of reports in which the disappearance of the *hali* system is attributed to increased mobility of the Dublas. Conversely, of course, the risk of losing money lessened the landlord's inclination to take on new *halis*. Another indication in this direction is that at one time the Anavils advocated the introduction of work-cards in order to prevent a farm servant from leaving his master without permission.[25]

The disintegration of the *hali* system can undoubtedly be interpreted in various ways. Some authors are inclined to put the stress on the inability or unwillingness of the Anavils to take farm servants into their employ. Others think it was the Dublas who, against the will of the landlords, withdrew from the traditional service relation. In both cases, however, the explanation of the disappearance of bondage is a purely economic one: a change in the situation which determined the demand for labor or, conversely, the need for work.

The available data show unmistakably that, economic changes aside, factors of a politico-social nature played a role. The traditional relationship between master and servant was increasingly undermined by external pressure. In the first place, the elements of coercion inherent in the *hali* system were no longer thought to conform with principles of good colonial government. Although nineteenth-century reports never glossed over the lack of freedom of the *halis*, the judgment of administrators in the present century was more harsh. They compared the *halis* to pre-Civil War plantation

25. Memorandum by Rao Saheb B. M. Desai to the *Royal Commission on Agriculture in India*, 1927, vol. II, pt. 2, 577, 601.

slaves in the southern United States.[26] In the state of Baroda in 1923, it was announced that to force laborers who were indebted to a landowner to perform labor in compensation was prohibited.[27] As for the British part of south Gujarat, the *hali* system was never prohibited there officially, but local officials were forbidden long ago to assist in tracing runaway farm servants. We should of course not delude ourselves that this regulation was carried out. In practice the Anavils could always count on the help of the lower government officials. It is therefore not surprising that an official prohibition such as that issued by the state of Baroda was not very effective either.

A man refusing to give his work, for which he has taken a loan is sometimes improperly prosecuted for cheating; and I regret to say that magistrates have sometimes taken a strangely perverted view of such prosecutions.[28]

Yet the fact that the landlord could no longer legitimately appeal to the local authorities when a farm servant had deserted him is of some importance. As I explained in chapter 3, the bureaucratic apparatus gradually became a little less susceptible to the interests of the Anavils. Thanks to the recent centralization of administrative decision-making on the district level or higher, the arbitrary activity of lower government officials—themselves often Anavil Brahmans—was increasingly weakened. While, formerly, to favor their own relatives and fellow caste members was a perquisite of their authorized position, Anavils who now wished to take advantage of their official standing for this purpose were guilty of unlawful actions.[29] As landlords the members of this caste remained dominant in the region, but the increasing bureaucratization of local authority inevitably led to a lessening of their authority. Those who had acquired a position in the government apparatus now had fewer opportunities to give priority to the interests of their own group than they had fifty years earlier.

The diminishing influence of the Anavil Brahmans is, for instance, apparent from their vain attempts to persuade the government to take measures against runaway *halis*. Their spokesmen ar-

26. *Census of India*, 1921, vol. III, pt. 1, 220.
27. *Census of India*, 1931, vol. XIX, pt. 1, 255.
28. Symington, n.d., 50.
29. "Several cases have been recorded in which the Dhaniamo employed the agency of the Government under the old regime for the purpose. It is only recently that the farmer's position has weakened in this regard." *Report of the Hali Labour Enquiry Committee*, 24.

gued that legal sanctions were needed in order to safeguard agrarian production. This attempt to curtail the mobility of the agricultural laborers was no more successful than the landlords' proposal that requests by agricultural laborers for a wage increase should be made an indictable offense.[30]

The attitude of the colonial government towards the *hali* system was also influenced to some extent by the pressure exerted by the independence movement. Gandhi, who was exceedingly familiar with the situation in south Gujarat, was one of the first national leaders to condemn *halipratha* in no uncertain terms. During his stay in the region he repeatedly declared that a struggle against colonial suppression was justified only if it put an end to lack of freedom in indigenous society, such as was suffered by the Dubla *halis*. The denunciation of the traditional service relationship, repeated explicitly in a report of the land-reform committee of the Congress Party in 1951, was a setback for the landlords.[31]

Partly in consequence of the Gandhian social movement, *halis* no longer acquiesced in bad treatment, but left their masters for good.

> The Dublas are undoubtedly becoming more independent, and more capable of combining in defence of their rights or interests. If they think they are stinted of their food ration or not paid a fair wage, they will desert the village in a body, as happened this year in Afva, Vankaner, Timberva, and other places.[32]

It should be added that increased contacts with the towns and the possibility of finding work outside agriculture if desired has made for a more independent attitude among agricultural workers.

Such reports show that those laborers rebelled against their bondage, and suggest that more and more Dublas preferred the freedom enjoyed by day-laborers.

5.3 A Supplementary Explanation

Whether the changes in relationships are primarily attributed to landlords or to agricultural laborers, or whether they are connected with economic or politico-social factors, the common denominator of most of these views is that they portray the disintegration of the *hali* system in terms of a changing *labor* relationship: a changeover from the bondage of servitude to the freedom of the day-laborers'

30. *Baroda Economic Development Committee, 1918–1919*, 1920, 131–32; Dave, 1946, 33.

31. *Report of the Congress Agrarian Reforms Committee*, 1951, 128–30.

32. *Report of the Special Enquiry . . . Bardoli and Chorasi Talukas*, 1929, 15.

existence. Although this representation is not wholly incorrect in itself, it is far from complete.

In the preceding chapter I distinguished another dimension of the *hali* system. Servitude is more than a labor relationship characterized by lack of freedom. The relationship between the *hali* and his master has been explained by me partly in terms of patronage. There is reason to assume that the patron-client dimension in the relationship between landlord and agricultural laborer has lost much of its importance since the beginning of this century.

This background should figure largely in any explanation of the disappearance of the *hali* system, and it is an aspect that emerges clearly from the information on changes in the relations between landowners and agricultural laborers elsewhere in India in recent decades. Srinivas and Bailey even say explicitly that the elements of patronage ceased to exist when, in the villages, traditional servitude fell into disfavor with the rise of a market economy.[33] Similar observations are made by other sociologists who have remarked on the loss of the close tie that had bound landless families to landowning families for generations.[34] The vertical pattern of relationships that was characteristic of the traditional situation was replaced by mutually antagonistic attitudes. Landowners on the one hand, and agricultural laborers on the other, have become parties with conflicting interests, and contacts between them increasingly reflect that situation. This process of change was closely related to the disruption of the stationary village economy and a break-through to an expanding economy on capitalist principles. Improved communications with the outside world, along with increasing production for the market, resulted in an enlargement of scale that put an end to the more or less closed order of the traditional rural community. The relationships between the various categories which it comprised were organized on new lines. In the various regions of India this process of course took place at different times, but in general the turn of the century is considered the point of beginning, and the First World War and the period after Independence as phases of acceleration. On the basis of my own investigation I shall set forth in the following chapters the course taken by this process in south Gujarat, as reflected by the changes in relationships between landlords and agricutural laborers.

33. Srinivas, 1955, 27; Bailey, 1957, 143–44.

34. See, for instance, Gough, 1955 and 1960; Cohn, 1955; Neale, 1962, ch. 9; Sivertsen, 1963; Béteille, 1965, ch. 5; Orenstein, 1965; and Harper, 1968.

PART III

Chapter 6

Present Relationships between Landlords and Agricultural Laborers in South Gujarat

6.1 Plan of the Field Work

An analysis of the relationships between landowners and agricultural laborers after the disappearance of the *hali* system was, in brief, the starting-point of my field work in 1962–63. There seemed no doubt that the *hali* system no longer existed. "There are no *halis* any more, they belong to the past," I was told at the highest administrative level when I paid a courtesy call there and explained the purpose of my stay.

This remark by an official was intended as more than a statement of fact. The word *hali* has ominous overtones in official ears, for its connotations with bondage—for many people even with a kind of slavery—are inconsistent with the image of India after Independence. It seemed not only obvious but also reasonable that officials preferred to speak of the *hali* system in the past tense. In various publications during the past few decades its imminent dissolution was indeed announced. The reliable study by V. H. Joshi on the process of social and economic change in a south Gujarat village, moreover, confirms this assumption. Joshi arrives at the conclusion that the *hali* system no longer exists. He describes a situation in which the agricultural laborers dispose freely of their labor; as a rule they work as day-laborers for one Anavil landowner or another.[1]

In short, these conclusions seem to confirm that the repeated predictions have at last come true. Formerly bound farm servants, the Dublas have now become free day-laborers.

Yet when I began orienting myself in the area of research in preparation for field work, I soon learned that this view is not en-

1. V. H. Joshi, 1957, 53–64.

tirely balanced. It is true that some of the agricultural laborers do not work for the same landowner day after day. Aside from these casual laborers, however, there is still a category, differing widely in numerical strength and depending on the locality, that consists of farm servants. On inquiry they turned out to have been employed for years, sometimes all their lives, by the same landowner. And I could not escape the impression that their labor agreement paralleled strikingly that of the *halis* of the past.

Should I conclude that the traditional service relation still existed, though to a differing extent in different places? This was indeed the opinion of a number of local political leaders and social workers whom I interviewed about it. Nevertheless their view may well lack discrimination. As I soon discovered, the nature of the farm servant's relation to the landlord has changed so drastically that it would be erroneous nowadays to call him a *hali*.

To begin with, I sought the changes especially in the disappearance of the unfree elements that had been inherent in the service relationship. In other words, certain Dublas are still servants but are no longer bound. But I was quickly compelled to abandon this point of view. The bondage of the agricultural laborers has not wholly ended. To suggest a transition from bondage to freedom is to give an incomplete picture of the process of change that has taken place. In the course of my field work I arrived at the conviction that this approach does not cover all the aspects and is even somewhat incorrect. Gradually I shifted the emphasis to the elimination of the element of patronage in the relationship between landlords and agricultural laborers in south Gujarat. Particularly in this sense, I believe, the term of *halipratha* no longer applies.

Thus, at the beginning of the field work proper I was burdened with a confused mass of contradictory information, and, being unfamiliar at the time with the past history of the situation, I opted for an open statement of the problem, which I formulated as follows:

—How is the disappearance of the *hali* system to be explained?
—What relationship between landlords and agricultural laborers has replaced it?

Without knowing clearly what I would find, I could in this way express my conviction that a good understanding of the pattern of relationships that has come into being required the gathering of data on both the traditional service relationship and the manner in which it changed during the past few decades. As I have provided a detailed description of the past situation in part II, I shall concentrate on present conditions in the report of my village study.

The way in which I stated the problem further implied that I should give close attention to the change in those contacts which are determined by the work situation: the recruitment of labor past and present, the varying conditions of employment of agricultural laborers by landowners, the duration of the service relationship, etc. But any shifts in this pattern of economic interaction, in the labor system, reflect a process of change which reaches much farther: a change in the relationship of Anavil Brahmans and Dublas, the wealthiest and poorest groups in the rural society.

6.2 Method

Before venturing to prepare a plan of research, I spent a month in a centrally situated, randomly chosen village in the plain of south Gujarat to get my bearings and to improve my command of the language. During this period I could familiarize myself with the field-work situation and accustom myself to local conditions. The advantages were many—for instance, I need not fear any too drastic consequences of my investigation proper if I should trespass in some way, and in particular if I should infringe any rules, an error which every stranger in a caste society seems to commit sooner or later. That initial stage was especially valuabe because it enabled me to obtain a better insight into the subject of inquiry which I had formulated in a distant country with no knowledge of the local situation. This first and as yet sketchy orientation was much more effective and profitable than a study of the literature during the same short period could have been. Traveling throughout the area, I was hospitably received in a great many villages; I talked to numerous informants, such as civil servants, "social workers," landowners, and agricultural laborers. During this phase of reconnoitering I decided on the definitive plan of my investigation and formulated the basic questions with which I began my field work. This first orientation, finally, prepared the way for the work and enabled me to select on less arbitrary grounds the two villages in which I was going to stay longer to collect data.

My subject of study was the changes in the relationships between landlords and agricultural laborers. From the outset I intended to select villages where part of the population consists of Anavil Brahmans and Dublas. For it was principally the members of these castes who were looked upon as the elements in the *hali* system, and it is they who still form the most important groups of landowners and agricultural laborers, respectively, in south Gujarat. But the two castes are not represented everywhere in equal strength. The Anavil Brahmans live chiefly in the *talukas* farthest to the south, hence the northern part of the plain did not primarily qualify for field work. Although there, as elsewhere, the majority of the agricultural laborers are Dublas, the landowners mostly belong to the caste of the Kanbi Patidars. To be quite clear, in the region to which the field work was largely confined there were other landowners besides Anavil Brahmans, whereas the agricultural laborers were not all Dublas. I concentrated my attention especially on those two castes, however, because the aim of the study was not to analyze the relationship between all employers and employees in agriculture but principally to trace changes in the relations between Anavil landlords and Dubla agricultural laborers.

From conversations with informants during the period of orientation, the fact emerged time after time that, aside from the social position of the Anavils as landlords, their relationship to the Dublas depends to a great extent on the following circumstances:

—The nature of the crop system: When agricultural operations are restricted to a few months of the years, the landlords prefer to employ casual labor. The use of permanent farm servants, on the other hand, is more feasible in areas of perennial agriculture.

—The extent of geographic isolation: If urban centers and the railroad in the west are nearby and easy to reach, the agricultural laborer can more easily find work outside the village. The farther from the city and the less convenient the transportation, the smaller his chance of alternative employment. Geographic isolation thus retards the disruption of the traditional relationship between master and servant, as I was told several times.

—The political background: It was suggested that in the British part of south Gujarat the *hali* system fell into disuse earlier because of the activity of the Congress Party and especially through the influence of Gandhi and his followers. In the territory of the principality of Baroda the process of change was much slower. There the Anavils had full scope and obstructed the emancipation of the Dublas.

I made the study a comparative one by dividing the field work between two villages that differed in the above respects. These factors, which predicted different patterns of relationship between Anavil landlords and Dubla agricultural laborers, were not used as a hypothesis, that is, they were not carefully formulated in a preliminary stage and verified by field work afterwards. They simply served as a lead for me to make a selection for research on the village level.

The choice finally fell on Chikhligam and Gandevigam, two villages which I named after the *talukas* in which they are situated and which are comprised in the present Bulsar district. To arrive at a decision I proceeded as follows. On the basis of data given in the district handbook, I listed a number of villages which at first sight seemed suitable for field work.[2] I visited some of these villages myself and made inquiries as to the social and economic background of others from local authorities. In continuous consultation with Professors I. P. Desai and M. B. Desai in Baroda I finally made my choice, in which an important consideration was also the possibility of separate accommodation for me.

With regard to the three factors outlined above, the two villages that were selected present contrasting aspects, which enabled me to carry out my plan of establishing, in the course of the investigation, the weight of each of those factors separately and explaining any difference in the results in this way.

Labor demand in Chikhligam is greatest in the periods before and after the rainy season. The winter months are the slack period in agriculture. Gandevigam, situated in a region called the garden of south Gujarat, has irrigation agriculture. The crop system there requires continuous employment of labor. The yield of agricultural production in this village is much higher than in Chikhligam and the produce is very largely sold on the market. Chikhligam is difficult to reach, and alternative nearby employment is completely lacking.

2. *Surat District Census Handbook*, 1951. At the time of my stay, the district handbook based on the 1961 census was not yet available. For the social-research worker the handbook is an excellent guide, providing various facts on every village: total number of houses and households, number of inhabitants divided according to men and women, number of illiterates, working population in and outside agriculture, the nearest railroad station and market place, and finally the facilities available in the village, such as school, cooperative, etc. As others discovered before me, however, the data given are not wholly reliable.

Gandevigam is on an asphalt road within easy reach of some towns containing various industries. Finally, in former times Chikhligam was under British rule, whereas Gandevigam belonged to the principality of Baroda.

In both villages — and this was the first criterion of selection — Anavils and Dublas are the leading categories of landowners and agricultural laborers. Labor demand varies not only with the nature of the crop system but with the social position of the landowners. The category of Anavil Brahmans which is regarded as higher, the Desais, have always been landlords. Some of the Bhathelas, on the other hand, still work on the land themselves, although the number of active cultivators among them has sharply declined in recent years. The greater differentiation within the latter group makes it possible to trace the influence of the landowner's status, which is determined by the amount of physical work he does on the land, and on the nature of his relationships with agricultural laborers. It is for this reason, not primarily because they form the majority of Anavil Brahmans, that I concentrated my research especially on the Bhathela group. Yet in this respect, too, there is a difference between Gandevigam and Chikhligam. Like the other garden villages in this region, the former is typically a Bhathela village, but Chikhligam is situated in a markedly Desai area, and one of the Anavil families in the village belongs to this category. The Desai style of life, in other words, has always been an example which could be observed at close range in Chikhligam.

Both Chikhligam and Gandevigam have in common with the other villages, which I considered in the first instance for field work, that the population of each was said to be less than 1,500. The division of the field work between two villages precluded my staying much longer than three months in each. By selecting two villages that were not very large, I hoped to avoid as much as possible the difficulties incident to the short duration of each stay. I am well aware of the limitations imposed on the investigator by a brief sojourn. I had no time, for instance, to interview systematically any village inhabitants other than Anavils and Dublas on the subject of the relationship pattern between the members of these castes of landlords and agricultural laborers. And I am aware that other such gaps exist.

Finally, I used my time as follows: The first phase was spent, between October 1962 and January 1963, in the city of Baroda and in south Gujarat, where I stayed in a pilot village for an experimental period. Early in February I installed my family, who had arrived from the Netherlands, in Baroda. The greater part of the field work proper took place between February and July. During the rainy season, from July to September, I was occupied mainly with literature study in Baroda and Bombay. I resumed the field work in September and concluded local research on the village level at the end of November.

6.3 Procedure

"May I see your questionnaire?" I was often asked when I explained the aim of my investigation in a university or the office of a government body. This question is in accordance with the marked preference in India for numerical data which are collected in great quantities during fleeting contacts with respondents. As mentioned in the preceding paragraph, I used the research technique of participant observation in attempting to gain an insight into the relationship between landlords and agricultural laborers in two villages in south Gujarat. At the end of my stay I verified my findings by a brief investigation in a third village on which I shall report in the final chapter.

Three months' field work in each of the two villages is scarcely adequate for participant observation. It could be said that the shorter the period spent in the village by the investigator, the more emphasis is laid on interviewing rather than on observing. It is true that I confined my attention in the main to the Anavils and Dublas, but the presence of other castes partly determined their relationship with each other. A minimum of knowledge of the whole village situation was therefore indispensable.

The central question was: What type of relationship has replaced the *hali* system? I tried to find the answer by collecting data concerning contacts between them from all Anavils and Dublas in Chikhligam and Gandevigam. I interviewed the members of the Anavil caste as to the size and composition of their property in land, the nature of their crops, the number of laborers in permanent service, their need of day-laborers, the wages they paid, working hours, and the like; insofar as they themselves or any members of their households were employed outside agriculture I questioned them concerning their work and education. The Dublas, on the other hand, provided me with information on the nature of their service status; for whom and for how long they had worked as farm servants or whether they were day-laborers, their wage and indebtedness, the daily routine of an agricultural laborer, which members of their households also worked on the land or in the master's house, and possibly what work they did, temporarily or permanently, other than agricultural. I have not wholly succeeded in quantifying my data on the economic situation and style of life of the two castes, owing to the short duration of my stay as well as to my impression that I had not been able to put into figures all the relevant facts needed for a really reliable report.

Thus I tried in vain to establish the average indebtedness—a wage component, as I will explain—of the agricultural laborers. The farm servants sometimes could not (would not?) tell me even approximately how much they owed and by how much the debt had increased in the course of time, while the amounts stated by the landlords were certainly too high.

A budget survey had to be omitted, because I could obtain only an incomplete picture of some important components of receipts and expenditure and their distribution throughout a year. The distortion caused by the sharp seasonal fluctuations can be corrected only by continuous registration. It is even more difficult to calculate the income from theft of crops or the expenditure on illegally distilled spirits, concerning which, understandably, the Dublas are not very communicative.

Nor have I been able to state with precision the frequency of running away among farm servants. The landlords' statements on this point turned out to be exceedingly unreliabe, yet I could not ask the Dublas who had left the village the reason of their departure. The Dublas who had come to live in the village were sometimes apt to conceal their motive for coming, but I could by no means assume that in all cases they had run away from a master in another village.

The condition of pauperism in which the Dublas live does not easily lend itself to the collection of numerical data. Many of them reacted vaguely and evasively to concrete and exact questions, because of ignorance, indifference, and suspicion.

Finally, too, the lack of stability of the unit of co-residence, another phenomenon of poverty, was responsible for the respondents' inability to give exact replies to questions. They sometimes had to remain silent when I asked after a husband, children, or parents because they did not know their whereabouts or present circumstances. Then, too, facts are concealed for completely different reasons. I did not succeed, for instance, in proving numerically that only a very small proportion of the Dubla children attend school. I was readily given access to the list of registered children, which contained a great many Dubla names. In conversation, the teacher did not hesitate to admit that most of them permanenty stayed away. But my attempts to take samples were resented and had to be abandoned.

Instead of suggesting an exactness that would be unfounded I have used the material I collected to give a broad outline and sketch some tendencies. A further quantification is certainly possible but requires a different method from the one I used.

My field work was appreciably influenced by the antagonism that exists between Anavil Brahmans and Dublas. It was not difficult to arrive at a good understanding with the members of the dominant caste. I was introduced into the villages by them and in their houses. They regarded it as evident that I would stay in their village quarter and that they were my obvious, even my only informants. I also found it easy to associate socially with them, not least because the English of some of them was better than my Gujarati. Especially at the beginning of my stay I spent most of the day in their company, and in fact this was expected of me. During that period I was able to collect material relating to the whole village system and to the Anavils themselves. The information they gave concerning other castes was generally very superficial and, moreover, not very reliable. The strikingly unfavorable picture, for instance, that the Anavils gave me of the behavior and way of life of Dublas was of interest as evidence of their dissatisfaction with the attitude of these subordinates but turned out to be completely incorrect in various respects.

Whereas the Anavil Brahmans were relatively easy to approach, it was much more difficult for me to build up a confidential relationship with the Dublas. I had to use some tact to overcome the resistance of the landlords of the village, who were not very pleased with my association with agricultural laborers. "You should ask us what you want to know," they gave me to understand. I made as many allowances as I could for their susceptibility, but ultimately I went my own way.

A much more pressing problem was the suspicion exhibited by the agricultural laborers as to my intentions. They identified me with the dominant caste, which was understandable enough.[3] I stayed in the midst of the Anavil Brahmans, which made a close connection with them inevitable. In their quarter I had been accommodated in an empty house. The Dublas met me for the first time in the houses of their masters, where I had been invited as a guest, or on the fields in the company of their landlords, who were accustomed to speak of the Dublas in the rudest and most offensive terms even if a servant or maid was present, with me as a silent observer.

The much greater distance that separated me from the Dublas may explain the reticence and suspicion with which they reacted to my first attempt at approaching them. For me this was the tensest period of my stay in the village. Fortunately, like the rest of the inhabitants the Dublas knew I was an outsider after all, as by my attitude and behavior I did not correspond with the picture of a high-caste Hindu: for instance, I had no objection to drinking illegally distilled spirits with them and tried to associate with them on a footing of equality. The friendship I had formed with a Mochi boy in Chikhligam, and with a Moslem boy from a village adjoining Gandevigam, stood me in good stead in this phase of the field work. Thanks to them the communication barrier with the Dublas was broken through for me—my Gujarati not being up to the dialect of this tribal caste—because, being members of the less favored groups in the villages, the boys were acceptable to the agricultural laborers as my interpreters.

Contacts with the Dublas multiplied as my stay in the village continued. The evident vexation this aroused in some of the Anavils, who regarded my behavior as improper, enhanced my reliability in the eyes of their subordinates. Answering my questions, however, continued to make many Dublas uneasy. After all, to give information about one's master was to commit oneself. To allay their suspicion I therefore rarely made notes on the spot of the conversations I had with them.

Usually I managed to keep out of the quarrels between Anavils and Dublas. If I did become involved for some reason or other and my role might be in doubt, I never hesitated to take the side of the latter. The following occurrence is illustrative. One evening, while I was sitting with a Dubla in front of his hut, his master came to call him in angry tones because he had not turned up for

3. See Baks, and others, 1965, 171–73.

work that day. A dispute followed, and when the landlord, obviously expecting approval, drew me into it I said something to the effect that anyone might like to take a day off occasionally and told him straight out that I did not care for his remarks about the laziness and mendacity of Dublas in general. As a result, my relations with this Anavil lost much of their cordiality, but this was something I had to accept. Emotionally, any such rather noncommittal expression of sympathy with the agricultural laborers came easy for me; moreover, it was of the utmost importance for my investigation that I should remain on good terms with this not easily approachable group.

For the same reason I addressed myself directly to the agricultural laborers in the investigation I carried out in a third south Gujarat village for checking purposes. Considering the short time I could spend there, it was impossible for me to win the confidence of both the landlords and the laborers, and the greater usefulness of good contacts with the Dublas weighed more. Since I did not live in the village itself but at a short distance in the small town of the subdistrict, there was, moreover, no need to take much notice of the landlords' feelings. Instead of going to them first I went, from the first day onwards, straight to the Dubla village quarter. Thanks to an introduction which I had likewise acquired without resort to the landlords, I stood well with the Dublas from the beginning. But the local power elite were greatly offended. After a few days the village headman—seated on a horse, the only one in the village, to emphasize his dignity—came to the Dubla quarter and without giving me so much as a glance interrogated the Dublas in a surly tone about "him there": "who is he?" and "what is he after?" This was wholly unnecessary as the landlords certainly knew all about my conversations with their subordinates. Partly to prevent them from venting their annoyance with my actions on their servants, I tried towards the end of my stay to get into touch with the landlords after all. I was accorded an icy reception. Some did not wish to receive me at all, and sent word from the inner recesses of the house that they were not at home.

The atmosphere of friction and conflict prevailing between landlords and agricultural laborers unquestionably made my field work more difficult. By including both these groups in the investigation, however, I could arm myself against partiality and avoid becoming too easily accustomed to caste cohesion and group judgment. A further advantage was the possibility of checking the reliability and completeness of information. Many of the discrepancies between the statements of Anavils and Dublas turned out, on further inquiry, to be due to a difference in emphasis. Widely disparate statements on the agricultural laborers' working day, for instance, were explained by the fact that Dublas measured it by the working hours at the peak of the season, whereas Anavils thought in terms of a normal day in the slack period. In this way I was able to correct the distortion that inevitably arises when one gathers information from a single caste.

To the extent that a subject of inquiry is socially more delicate, and certainly in a situation of conflict, the chance increases, especially when participant observation is the method of investigation, that the fieldworker has to take sides. It is a choice he may feel obliged to make or to which he may be compelled by his surroundings. In the course of my own field work I never made a secret of my own stand when I was explicitly asked about it. This was one reason why I endeavored to proceed as objectively as possible in collecting the data comprised in the following chapters.

Chapter 7

Chikhligam

My Subodbhai is a vakil,
And Vinodbhai is a doctor,
Ramanbhai begins his studies
With the English A, B, C.
T. B. Naik, *Songs of the Anavils of Gujarat*

O mother, I can't dehusk the grain,
I can't grind the corn,
I want to go away to Bombay.
To hell with your money,
To hell with your bride-price,
My married one lives in an old tottering hut.
Mother, I want to go away to Bombay,
I don't want your drinks,
Nor do I want your food,
Mother, I want to go away to Bombay.
O mother, to Bombay
I would have gone by train.
O mother, in Bombay
I would have sat at [a] window and seen through it.
O mother, in Bombay
I would have eaten rotli *with tea*
And fine rice too.
In Bombay, O mother,
I would have walked on a footpath.
I want to go away to Bombay.
T. B. Naik, *Dubla Songs*

7.1 Region, Village, and Population

Chikhli *taluka* is situated in the transitional area between the densely populated and fertile western plain and the relatively inhospitable and infertile eastern part of south Gujarat. The local town, with over 5,000 inhabitants, is a trade and civic center for the villages in the subdistrict. Here are the local government offices, some doctors and barristers, and the traders and shopkeepers on whom the villagers of the region depend for the sale of their agrarian products and the purchase of consumer goods. Industrial activity is limited to a number of workshops, each with a few employees. As there are bus connections with the urban nuclei of the surrounding subdistricts and with the railway line in the west, the town is a link in the contacts of the rural population with the outside world. The national highway from Bombay to Ahmedabad also runs through this regional center; it is mainly important for through traffic.

Chikhligam is about four miles from the *taluka* town. Rough country roads, running through terrain that is dry and arid both in summer and in winter, connect the village with those surrounding it and with the highway at two miles' distance. From that point, the inhabitants of Chikhligam can go to town by bus four times a day, but they rarely use this means of transport. In the wet season, when the country roads become impassable and the village is practically isolated from the outside world, the villagers can get to town only along the highway, which is normally not used except by those who own a bicycle. Two inhabitants subscribe to a regional newspaper, and there is a single radio set in the village, which was out of order during my stay. There is no running water, electricity, or any other modern convenience.

On the way to the village, the visitor passes through undulating terrain. The high, dry ground (*jarayat*) is stony and uneven, and dotted with *babul* trees (*Acacia arabica*). This *jarayat*, which in Chikhligam constitutes two-thirds of the total village land, is of inferior quality throughout the subdistrict and not very suitable for agriculture. The depressions in the terrain, on the other hand, have been leveled and are surrounded by embankments. This is the *kyari* soil on which in the wet season the rain water is collected for rice cultivation. Along a small river, which forms the eastern boundary of the village area, there is, finally, the very fertile *bagayat*, consisting of irrigated garden plots.

The returns of the 1951 census—on which my selection of field villages was based—had made me expect a village population of a little more than 1,000, which was well within the limits I had set. But whereas in 1951 only 900 inhabitants were registered, their number had risen to over 1,800 at the 1961 census. The explanation of this apparent doubling is the seasonal migration of a large proportion of the inhabitants of Chikhligam during about seven months of the year. In 1951, Chikhligam probably already had about 1,500 inhabitants, but the census of that year was taken in a period when the migratory workers had not yet returned to the village (see table 7.1.1).

TABLE 7.1.1

DISTRIBUTION OF THE INHABITANTS OF CHIKHLIGAM, ACCORDING TO CASTE*

	Number
High castes	
Anavil Brahman (Landlord)	125
Bania (Merchant)	5
Intermediate castes	
Darji (Tailor); Ghanchi (Oil presser); Valand (Barber); Mochi (Shoemaker); Vairagi (*saddhu*)	75
Low castes	
Dhodia (Petty cultivator)	1,360
Dubla (Agricultural laborer)	200
Untouchable castes	
Garuda (Priest); Dhed (Weaver)	90
Moslems	
Shopkeeper; Barber	5

* These figures, which I obtained from the village clerk, are only approximately reliable. I have been able to verify only the figures for Anavils and Dublas.

Despite the large number of inhabitants I persisted in my choice of Chikhligam. The members of the tribal Dhodia caste, who form three-quarters of the total village population, live outside the village nucleus in ten hamlets scattered over the whole village area. Each of these *phalias* comprises ten to twenty houses, whose inhabitants are mutually related by patrilocal residence. The hamlets, which are connected by narrow paths, are surrounded by the fields belonging to the inhabitants. Not only are the Dhodias spatially separated from the rest of the village population, but apparently they do not depend much on other castes for their needs. Their economic and social life is set within their own *phalia.* Each of the two largest *phalias* even has its own shop, with a modest stock of goods. Nor do the Dhodias mix with the other inhabitants at village festivals, as I observed when the bonfire was lighted on the occasion of *holi.*

Under those circumstances, I could easily restrict my field work to the nuclear village, in which the members of the other castes live, with Anavils and Dublas as the largest groupings. A pond, a large public well, a small school building, a decayed *dharmasala* (guest house), and a shabby little mud temple constitute the village center. As a center, however, it is especially recognizable because the Anavils live here. The members of this dominant caste tolerate no other neighbors but a Bania family in their street. Most Anavils live in large brick houses containing a series of three spacious rooms with a veranda in front, and the newly built houses in the street have an upper storey. The stable is separated from the living quarters by a large yard. Some Anavils have a private well behind the house.

A second street is the domain of the Mochi caste, at present still represented in the village by six households. The houses in the Mochi street are also of brick, but much smaller and lower, and a number of them are empty. Two are used as shops, one of which belongs to a Ghanchi family and the other to a Moslem shopkeeper, who is in the village only during the daytime. Most of the scattered houses of the three Darji households and the Valand have tiled roofs but mud walls.

The Moslem barber lives among the Dublas, whose block has a cluttered appearance. The huts in which they live are not built in rows, but in a rambling order. The Dublas' huts, the worst in the village, generally have only one room and are built of mud brick; their thatched roofs hang so low that one has to stoop to enter. There is no ventilation except by the open entrance. The total area of the hut is often so small that there is not enough sleeping space for all the members of the household. About six years before my stay, nearly half of the Dublas moved to a piece of ground that belongs to the neighboring village. Administratively they no longer live in Chikhligam, but because they still work temporarily or permanently for the Anavils of the village as agricultural laborers, I included this group in my investigation. Their dwellings in the new quarter, built with financial support from the government, are larger than their former huts and have tiled roofs. Finally, on the fringe of the village nucleus are to be found the houses of the untouchable Garudas and Dheds. Their surroundings look very well kept.

Little can be said with certainty about the age of the village. The Anavil Brahmans assert that their ancestors of several generations ago came here from villages farther to the west. As I noted in chapter 3, the *pax britannica* put an end to the original way of life of the tribal Dhodias who, as primitive cultivators, used to shift periodically to other grounds, returning after some years to their point of departure. It appears from the old administrative records that they were still very mobile at the beginning of the colonial era. Their sedentarization was promoted by the founding of villages in this area on the initiative of high-caste Hindus from the plains. The Anavils of previous generations, in particular, enjoyed a reputation as founders of villages. The origin of Chikhligam is probably similar to that of a nearby village described by Mukhtyar.[1]

This author states that immigrant Anavil Brahmans persuaded members of artisan and serving castes to settle in the newly reclaimed area. To this it should be added that, in this preponderantly tribal region, no fully formalized caste structure arose. Many local-level studies are reports of villages which the investigators selected precisely because of the presence of a large number of castes. In comparison with these, the caste composition of Chikhligam is relatively simple. No doubt this is connected with the relatively undeveloped functional interdependence of the inhabitants in the region. The Dhodias, who form a large part of the population of the subdistrict, have remained more or less autonomous as members of a caste of tribal origin.[2] Only on a limited scale do they use goods and services supplied by other castes, and even now they maintain few contacts with the other village inhabitants.

For the Anavil Brahmans, highly placed in the hierarchy, it seems to be much less easy to dispense with a large number of castes. As I said earlier, however, their Dubla farm servants are accustomed to perform numerous services for them — from water-bearing to

1. Mukhtyar, 1930, 44–45.

2. For an extensive description of the Dhodias, the reader is referred to the ethnographic study of this tribal caste by A. N. Solanky.

washing garments, cleaning the house and stable, etc. — which elsewhere in India are carried out by diverse castes to whom these tasks have been specifically allocated.

7.2 Economic Position and Activities of the Inhabitants

In an agrarian community the possession of land is naturally an important indication of position in the economic structure. Anavil Brahmans, less than 10 percent of the inhabitants, own almost as much land as the Dhodias, who form nearly three-quarters of the population. The land owned by an Anavil household is many times larger than that of other households, averaging 19.4 acres as against an average of, for instance, 2.2 acres per Dhodia household. Another indication of the ascendancy of the Anavils is that much the greater part of the coveted *bagayat* is in their hands. This concentration of their tenure in the arable land is a measure of their dominance.

It may be assumed that the Anavil Brahmans have laid claim to a disproportionately large part of the total holdings ever since their arrival. Informants from various castes expressed the belief that at the beginning of this century proportions were no different. If the members of the dominant caste deprived Dhodia farmers of their property rights, this must have happened in a more distant past; it is not due to a development which began only after the turn of the century. As early as 1865, in the first cadastral registration of Chikhli *taluka*, it is reported that

> The Desais have managed through the influence in the revenue administration of the pargana they have retained from the time of our predecessors in the country, and by pandering to the passion of these degraded creatures [the *adivasis*] for drink, treating them at the same time with a rude sort of kindness, to get possession of the richest lands.[3]

This quotation answers the question of how the titles came to change hands. Excessive drinking caused the tribal population to fall into debt to the newcomers and the Anavils, who in the villages represented the government, exploited their politically dominant position. They probably compelled the Dhodias to become sedentary by laying claim to part of the area in which they wandered as shifting cultivators. It was penetration from the west that made land a scarce

3. Cited from the original report in *Papers Relating to the Revision Survey Settlement of the Chikhli Taluka of the Surat Collectorate*, 1897, 62.

commodity, and the subjection of the autochthonous population to the incoming group was reinforced by taxation.

TABLE 7.2.1

CASTE MEMBERSHIP AND LANDHOLDING IN CHIKHLIGAM*

Caste	*Percent of village population*	*Property*		*Extent of gardens*	
		Acres	*Percent*	*Acres*	*Percent*
Dhodias	73	350	44	7	7.5
Dublas	10	33	4	4	4
Anavils	7	297	37	77	81.5
Harijans (Dheds and Garudas)	5	49	6	3	3
Mochis	3	47	6	—	—
Banias)	) 2	14	2	3.5	4
Outside ownership)	)	10	1	—	—
Total	100	800	100	94.5	100

* Figures provided by the village accountant.

The Dhodias are the small peasants of the village. Two processes led to the existing pattern of small landownership in this caste: first, the accumulation, as described above, of the most valuable soil in the hands of the Anavil Brahmans during the previous century, and secondly, fragmentation into scattered plots because of demographic growth in the more recent past. With a few exceptions, all Dhodia households own a 2 to 3 acre plot of land which mainly consists of not very fertile *jarayat.* The rapid increase of their numbers in the past few decades owing to the falling death rate has resulted in increasing pressure on the land. The labor of the members of each individual household usually suffices for their own plot. Apart from that, the Dhodias of the same hamlet occasionally work for each other reciprocally at harvest time. Formerly this exchange of labor (*sandhal*) was more widespread. Only a few of the larger farmers among them, who do not wish to supply labor on a footing of reciprocity, employ paid casual labor from time to time. Owing to resources other than those acquired through agriculture, some differentiation in economic position has gradually come about among the Dhodias. Income obtained from outside the village has enabled some of them to extend their holdings and to raise themselves above the level of their less fortunate fellow caste members. In former times many Dhodias were sharecroppers of Anavil Brahmans, and in that capacity they were obliged to assist the landlord on his land at the peak of the season. With the termination of nearly all those crop-sharing agreements by the Anavils some years ago, such field services provided by the Dhodias also came to an end.

Thanks to seasonal migration, the small cultivators are now less dependent on the dominant caste of the village. Some of them still occasionally perform agricultural labor for Anavils against payment, but they do it grudgingly and only if they need the money badly. With their own team of oxen, some Dhodias plow the land of those members of the dominant caste who keep no servant and, sometimes, no cattle.

Most of the land owned by Mochis and Harijans was bought some decades ago. With money earned in town, some members of this caste purchased plots in order to provide for the relatives they left behind. A few Dheds, whose ancestors had been entitled to the produce of certain plots in exchange for services rendered to members of the dominant castes, such as weaving garments and disposing of dead cattle, have now been given ownership of this land. Some Mochis who emigrated to East Africa bought land as an investment and leased it out to Anavils, Dhodias, or Dublas on a crop-sharing basis. In most cases they lost this land because of the tenancy reforms; only those who returned to the village kept their property. Like those of the Dhodias, the holdings of the families of these non-agrarian castes are so small that they do not need the regular labor of other workers, and it is beneath their dignity to work on the land of others.

The agricultural laborers of Chikhligam are mostly Dublas. They are, in large majority, completely landless and do not even own the piece of land on which their hut stands. That there are a few landholders among them does not signify that this caste originally owned more land and was gradually dispossessed. When asked about it, my informants told me that their plots, mainly consisting of infertile *jarayat*, had come into their possession at the beginning of this century. According to the Dublas, the land then belonged to no one, and the government had parceled it out to those interested—Mochis and Dublas. Informants claimed that the majority of the land now owned by Dublas had originally been the property of the Anavil caste, but had been relinquished about the turn of the century by the Anavil generation of the time. Some of the plots had belonged to an Anavil widow, who died the last of her family and left the land to her Dubla farm servants. This explains why one Dubla even owns a plot of the highly prized *bagayat*. Finally, after having worked for years in Bombay, two Dublas managed to acquire a plot of land with their savings. Actually the landed property of this tribal caste is even lower than appears from table 7.2.1. The Dublas who removed to the new houses are no longer registered in Chikhligam, although they still work there. All the landholders are among those covered by the table, i.e., those who remained behind. The Dubla agricultural laborers work almost exclusively for the Anavils of the village, in permanent service or on daily wages.

The economic positions of the members of the dominant caste also show considerable differences. In certain cases, Anavils have less land than the largest Dhodia cultivators. Even so, however, they are more favorably placed, for their land is of far better quality, and moreover they can count on the support of richer relatives. They clearly profit by their membership of the dominant group in the village. They can, for instance, borrow more money, and on more favorable terms, than the Dhodia farmers.

The Dublas, on the other hand, are not all landless, as was explained above. Six of those who call themselves cultivators own more land than the poorest Dhodias. In comparison with the other members of their caste, they are undoubtedly better off. They have one, sometimes two oxen, a cow, a couple of goats, and some simple agricultural implements. Their houses are larger than those of the landless Dublas, they own a greater variety of household goods, and they consume more, if not better, food. Nevertheless they are the poorest of the cultivators, and all of them carry out additional agricultural labor for the Anavils of the village. There is, in other words, a clear correlation between caste membership, property in land, and economic position. With a few exceptions, the Anavil Brahmans can be characterized as landlords and the Dublas as agricultural laborers.

Nevertheless, only for a small part of the village population is agriculture the sole means of existence. Despite the geographic isolation of their village, the majority of the inhabitants depend to some extent on income acquired elsewhere. Living in Chikhligam and employed in the village in other than agricultural work are: three shopkeepers (one Ghanchi, two Dhodias), three tailors (Darjis), two barbers (one Valand, one Moslem), one temple servant (Vairagi), one shoemaker (Mochi), two smiths-wainwrights (Dubla father and son), and one *panchayat* servant (Dubla). A few of the inhabitants work in surrounding villages or in the *taluka* town as schoolteachers (two Dhodias), postmen (two Dublas), counter-clerk (Anavil), shoemaker (Mochi), and *panchayat* servant (Dhed). Finally, seven Dhodias cycle each day to Bilimora, on the westbound railway line, where they work as factory hands. The Garudas in the village are priests and genealogists; they visit their clients, the untouchable Dheds, who live scattered everywhere in south Gujarat.

As these data show, nonagrarian employment in Chikhligam and its immediate surroundings is of little significance, less than 10 percent of the male working population being fully or partly employed in this way. However, the income from agriculture is supplemented in another manner. A great majority of the inhabitants of Chikhligam leave the village for shorter or longer periods, and some even stay away and return at most for a brief visit. Where they go and what they do is determined mainly by the caste to which they belong. The chances of finding work outside the village depend on migration traditions, on the nature of the work, and on the education of the worker, factors that differ for each caste.

The Anavil Brahmans are employed in the highly skilled and best-paid professions. Among the members of this caste who have left the village are a doctor, a college teacher, a stationmaster, an engineer, a manager in the state life insurance company, a moneylender, two garage owners, a mechanic, and a worker in a chemical factory. Five other Anavils are employed in government services. All live in towns, and all but one in south Gujarat. Many still own land in the village or are entitled to part of the produce of the family holding. Their agricultural interests are taken care of by relatives or have been entrusted to fellow caste members.

Most of the Mochis who have emigrated in the past have gone to East Africa, where they are engaged in their caste occupation or as shopkeepers. Others work in these trades in Bombay or Bilimora. The Dheds used to work as house servants for Europeans in Bombay and are now employed as such by well-to-do and westernized Indian families in that city. The Garudas were at no time economically attached to Chikhligam. In recent decades a number of them have gone to Bombay, where they practice their occupation of priests or are otherwise employed, for instance as drivers.

Relatively few Dhodias and Dublas have permanently left the village, apart from those who have gone to live in their wives' villages after marriage (*ghar jamai*). Usually, those who did go away are employed as unskilled laborers in Bombay, Surat, or the smaller railway towns in south Gujarat. The parents of many of them have not heard from their sons in years, and do not know where they are or what they are doing. Upon the death of relatives left behind, those who have gone away are sure to be forgotten. Their exact number is therefore hard to determine. According to my data, nineteen Dubla men from twelve huts are permanently employed outside the village.

Contacts of the tribal castes, small cultivators, and agricultural laborers with the city environment are as yet not very intensive. They have no specific knowledge that would equip them for nonagrarian employment, and being unskilled they qualify for only the least-favored and lowest-paid jobs.

For both Dhodias and Dublas, migratory labor is of far greater importance. Seven or eight months a year are spent by more than half of them in the brickyards near Bombay. The exodus begins after the wet season, but most of the migrants do not depart until the grass has been cut in mid-November. They stay away until the end of May or the beginning of June. The brickyards are small enterprises operated in the open air, and activities are concentrated around one or more kilns. The laborers are accommodated on the site in improvised huts of loosely piled bricks with roofs of thatch, branches, or leaves. Clay is brought in from the vicinity and made into bricks by teamwork. The *patla*, a gang of six laborers, molds the clay, forms the bricks and puts them out to dry. Because work is done in the open, it must be suspended during the monsoon. During the winter and summer months, only the aged and infirm, and occasionally women and children, remain in the village. Especially in the Dubla quarter, many houses are closed up for these periods. Complete families leave Chikhligam, and all—men, women, and children—work in the brickyards.

This form of migratory labor has increased considerably since the Second World War. Although a number of men of the older generation worked in the brickyards before the war, the demand for laborers in this branch was low, and many brickyards were functioning only part of the dry season. This occasionally occurs even now, so that the laborers must already return to Chikhligam in March. During and after the war, however, building activities in

Bombay and its surroundings expanded enormously. The number of brickyards increased accordingly, and their production was sharply accelerated.

The laborers are engaged by a labor-recruiter (*mukadam*), who is connected with one or more brickyard owners directly or through a contractor in Bombay. The recruiter often lives in the village, belongs to the same caste and has been a laborer himself. Insofar as the migrants are not picked up by trucks, the recruiter provides the money for the journey, and usually they all travel together to Bombay by train. The labor-recruiter is the link between village and brickyard. He goes with the laborers to Bombay, keeps accounts if he is literate, and pays the weekly allowances. Male laborers receive 6 rupees a week for living expenses, and women are given 4 rupees a week, but wages are not paid until the end of the season. By this system the laborers are attached to their place of work, and the brickyard owner does not run the risk of losing laborers halfway through the season. Accounts are settled upon their return to the village at the beginning of July, when the laborers receive what is owing to them. The amounts they save in this way vary widely, but even when the whole family works, they rarely have more than 100 rupees left, usually less. During the monsoon, when the laborers are in the village, they can get advances on their wages of the coming season through the labor-recruiter. In this way the brickyard owner is assured that he will have enough labor the following year, whereas the laborers, for whom there is not enough work on the land in the rainy season, can pay their daily expenses in the village.

This system makes for a lasting contact between the parties concerned. The laborers find employment waiting for them at a large distance from the village, and the brickyard owner constantly has all the labor he needs. On the other hand, whereas the party offering employment (the brickyard owner) has insured himself against nonfulfilment, the laborers have no redress against unjust treatment. The labor-recruiter receives 150 rupees for each gang delivered by him. If the gang does not appear at the beginning of the season he loses the premium and is, moreover, answerable for the advances that were paid through him to the members of the gang during the last wet season. His interests therefore coincide with those of the brickyard owner. Inasmuch as the wages are not settled until the end of the season, the laborers cannot leave the brickyard ahead of time. If they do, and if at the moment of leaving they are still indebted to the brickyard owner, as they often are at the beginning of the season, the other members of the gang are jointly answerable for their debts. Usually the laborers, nearly all illiterate, do not know exactly what amount is owing to them at the end of the season, so that the owner can give them less than they are entitled to. At the instance of the brickyard owners the labor-recruiters nowadays keep in touch with each other. Laborers who have run away, or who have taken up advances in the wet season without reporting for duty afterwards, are excluded, and occasionally civil actions are brought against them. Although it is a foregone conclusion that they cannot repay the money, such actions are used as means of intimidation. To organize the laborers into unions is difficult because of the seasonal character of the work. Moreover, the trade unions in Bombay regard these migrants as unwelcome intruders who depress the local wage level.

In an underdeveloped economy, the mechanism described above ought to balance wide local differences of demand and supply. But the laborers are bound hand and foot to the labor-recruiter, on whose collaboration they depend if they want to be employed again the following year. On the other hand, the power of the *mukadam* should not be overestimated. He lives in the same village and belongs to the same caste as the laborers. As a young man he usually has been away under the same conditions as those who now depend on him and who are sometimes related to him. Social pressure prevents him from totally identifying himself with the brickyard owner's interests or from systematically exploiting his contract laborers for his own ends.

Dhodias and Dublas participate in equal proportions in the yearly migration to Bombay. Yet there is an important difference in the economic position of the two castes. Their income from the brickyards has enabled the Dhodias to keep their family property, if only by progressive fragmentation, and to hold their own as petty cultivators. This also explains why more members of this caste remain in the village. It is the young people who go away, and only the women who have not been married long accompany them for a few years. Then they, too, remain behind to look after the cattle and harvest the winter crop, and their children go to school. The Dublas, on the other hand, having no land to provide for any relatives remaining in the village, must take their families with them. A Dubla can go to the brickyards only if the other members of his household work too, for earnings are so low that he cannot send money home. Migratory labor becomes impossible in cases of poor health or pregnancy of the wife.

The Dhodias' landownership, unremunerative in itself, puts them on a higher level than the landless Dublas. On the other hand, the Dublas continue to depend on agricultural labor because their income from seasonal migration is temporary; they thus remain at the disposal of the Anavil landlords of the village.

7.3 Caste Hierarchy and Political Structure

The settlement pattern and the villager's economic condition and activities indicate the caste to which he belongs. His position within the caste system appears from the pattern of social interaction. Norms regarding the accepting of food and water are indications of ritual status in the caste hierarchy. Thus, members of a high caste will in general not accept food from or eat together with those of lower castes. To some extent there is a relationship between a caste's economic and ritual status. This is true especially of the mutual allocation of position among the Anavil Brahmans, Dhodias and Dublas, the largest castes. However, the untouchable Dheds and Garudas, lowest in the hierarchy, are in better economic circumstances than the members of the tribal castes.

As elsewhere, actual behavior in Chikhligam deviates from the norm. In contradiction to caste endogamy, the Vairagi lives with the daughter of a Darji in the village. Although Anavils do not accept water from the untouchable Dheds, the village headman, a member of the dominant caste and a politically active member of the local section of the Congress Party, says he does accept water from Dheds in other villages. Again, the behavior pattern on which the place of the Anavils in the caste hierarchy is based includes vegetarianism and abstinence from alcohol, but some Anavils barely succeed in hiding the fact that they indulge in drinking illegally distilled spirits at the Ghanchi's, or worse, buying it from Dublas. It is an open secret that there are Anavils who commit the even greater sin of going to the dwelling of one of the Mochis to eat meat. In the eyes of their fellow caste members these Anavils have fallen very low because they associate with Mochis and Dublas, especially because they do it in the village itself.

These are individual deviations from the prevailing norm. Of more importance is the fact that many people of some castes leave

Chikhligam for years at a stretch and return only occasionally for short visits to relatives. Whereas interdependence to some extent implies the necessity of recognizing the relative positions of the castes, long absence may lessen commitment in the village community. Little is left to bind Mochis, Dheds, and Garudas to Chikhligam. Their children grow up in the village, but nearly all go away as soon as they are old enough, mainly to Bombay.[4] The Garudas occupied a marginal position in the village community even in the past when they all practiced their priesthood. The relatives they have left behind depend on the money sent from outside and have practically no contacts in the village.

A change in the pattern of a caste's economic activities may mean that its members are striving after a higher position in the traditional status hierarchy. When, however, many members of a caste leave the village for long periods to earn their living elsewhere they may also be withdrawing from the caste relationships in their village. The way in which the Mochis who emigrated to East Africa behave will differ profoundly from that of their fellow caste members who are still employed in the village economy. As for Dheds who work for well-to-do Indian or European families in Bombay, and who return to their native village for only a few days a year, dressed in trousers and nylon shirts and showing off their knowledge of English to the investigator, other norms and values obtain than those that were traditional for their parents, who were the weavers of the village and carried out a number of tasks regarded as defiling.

Seasonal migration, too, is bound to influence the relations between the villagers. For the Dublas this will be discussed below. In the Dhodias' case, the yearly migration has accentuated the economic and social isolation already inherent in their residential separation, and lessened their influence on the organization of village matters.

For the pattern of interaction, not only the extent of incorporation of the various castes in the village economy is of importance, but also their numerical proportions, which greatly influence intercaste relationships. Although it may be useful to know that, in the social hierarchy, a Ghanchi is higher than a Dhodia, this knowledge

4. I could not ascertain when members of these castes began leaving the village. Dheds from the nearby village of Atgam worked for Europeans in Bombay as early as the nineteen-twenties (Mukhtyar, 57). At the turn of the century, emigration from the *taluka* was still negligible. Presumably it has gathered momentum since then. Especially after the First World War, the population of Greater Bombay increased rapidly.

is not very relevant if three-quarters of the village inhabitants belong to the Dhodia caste, and some other castes are represented by only a single or a few households each. The arrangement of castes by ritual standards, therefore, does not reflect the actual distribution of power in the village, which is determined by the dominance of the Anavil Brahmans and the numerical preponderance of Dhodias.

The Anavil Brahmans have always been regarded as the people in authority in the subdistrict,[5] and their ascendancy continues today. Although this caste comprises less than 10 percent of the inhabitants, it has as many representatives in the village council as do the Dhodias, i.e., three. This *panchayat* meets only a few times a year. The daily course of events is wholly governed by the village headman, who has always come from the ranks of the Anavils. The police *patel*, who has to report all trespasses and crimes to the police in the *taluka* headquarters, is an Anavil. The local credit cooperative presents the same picture—most of its ninety-one members are Dhodias, but five of the seven members of the board, including the secretary, are Anavil Brahmans.

Anavil ascendancy is apparent in various ways aside from the fact that they occupy the formal positions of power. They act as arbiters in quarrels between members of other castes, and the leading Anavil faction has put a number of influential Dhodias under obligation, so that its power basis is safe.

The influence of the dominant caste is apparent, for instance, in the effect of tenancy reform in Chikhligam. The purpose of the Tenancy Act was to have the land revert to the actual cultivator, i.e., the tenant. In the past, landholding villagers who had left Chikhligam—mostly Anavils and Mochis, but also a few Banias and a Brahman from the *taluka* town who owned land in the village—used to lease out their property. When the new act became operative, the property rights of non-Anavil absentee owners shifted to their former tenants, mostly Anavils and Dhodias. The Mochis from Chikhligam who were in East Africa lost their *jarayat* to Dhodias and Dublas, and their *bagayat* to Anavils, who rented only the most valuable land. Most of the land leased in the village, however, belonged to Anavils. The income of one Desai family derived

5. Compare the report cited in fn. 3 above, pp. 62–63: "These Bhathelas are notorious as the greatest intriguers in the country. This character, assisted by the hereditary influence I have described them as possessing, has enabled them to worm themselves into positions of power as patels of villages, which they have used to the injury of the revenues, by exacting labour from the lower castes of cultivators, and thus preventing them to improve their own conditions."

almost wholly from land they had leased out, chiefly to Dhodias. As did the other Anavils, this family lost very little property in the tenancy reform. A good example is the case of a Dhodia who grew millet on *jarayat* he had rented from the village headman. He wished to have the land registered in his name, as he was entitled to do under the Tenancy Act. The village headman had been able to prevent this, as he himself said, by exerting some pressure. Although for him, in his function, this sort of thing is very easy, the same is true to some extent for all his fellow caste members. Some of them still lease out land illegally, but without anxiety, to Dhodia cultivators on a crop-sharing basis. The Anavils think this land reform unfair. It is not right to take away a man's property, they say. In regard to an Anavil who had told me indignantly how a Dhodia had cheated a Mochi owner by "stealing" land from him in this way, I was told later that he had "received" a *bagayat* lot from the same Mochi under the Tenancy Act. Not only did the members of the dominant caste not lose any property, they added to it. Some Dhodias and Dublas managed to become the owners of the land they had rented from Mochis living in East Africa and from a number of outsiders, but not from Anavils living in or outside the village.

The exercise of power by the Anavil Brahmans might in theory be counterbalanced by the numerical strength of the Dhodias. But, as has become clear, the latter's weak economic position makes them extremely vulnerable. It is therefore not surprising that Anavils speak kindly of Dhodias — a mild-tempered caste who do as they are told. The landlords can harass them in various ways, for instance by denying credit to "troublesome" and "unwilling" Dhodias, either directly by not lending them money individually, or indirectly by using the Anavils' majority vote in the board of the credit cooperative. Contacts with the government often run via the influential members of the dominant caste, and Dhodias who know how to behave can be sure that their business with the government will receive proper attention. A frequently used device in the "dry" state of Gujarat is to prosecute unruly Dhodias for illegally distilling and drinking alcohol, that favorite pastime of all castes of tribal origin in the region.

The following occurrence is typical of the behavior of the Anavils. One night during my stay in Chikhigam, police from the *taluka* town raided some Dhodia hamlets. They seized three men, whom the police had already tried to apprehend by day several times but who had always managed to evade them. They were wanted for running away from a brickyard while owing money to the owner. Ostensibly, however, they were taken into custody for distilling and drinking alcohol, an offense for which practically all members of the caste could have been arrested. Distilling equipment was found in the houses of two of the prisoners. The Anavil police *patel* had been informed of the time of the raid, and had chosen some fellow caste members, one of whom was a notorious drinker, to assist the police. Early in the morning the raiding party returned to the Anavil street in high spirits, and spent the rest of the day exchanging stories about the successful expedition. Relatives of the arrested Dhodias came to ask the Anavils to put in a word for them with the police; it was harvest time, and the men could not be spared. They were told that, first of all, two Dhodias who had escaped the night's raid because they had been fishing on the river must give themselves up. Shortly afterwards the two did come to report to the police *patel.* As for the others, the Anavils said, it had been their own fault, and if they had done wrong they must pay for it. To my question why certain Dublas had not been arrested for the same offense that night, the *patel* answered indignantly that these were their own laborers and they could not possibly spare them.

Nevertheless the power structure is beginning to change. Because so many of them go to work at the brickyards, the Dhodias are no longer quite so poor as before, nor so completely bound economically to the Anavils. Apart from renting land from Anavils, many Dhodias used to perform field service for their landlords, but now only the poorest of them do this. More and more of their children attend school. The Anavils were shocked when I figured out for them that, in all, more Dhodia children from Chikhligam (ten) go to secondary school in a nearby larger village than Anavil children (six). The son of a labor-recruiter from the village is at college in Bilimora.

The strengthening of the position of the Dhodias has been apparent, too, in the siting of a new building for the lower classes of the village school. On the Dhodias' request, this building was not put near the Anavil street, but halfway between the nuclear village and some Dhodia hamlets, so that the children of both castes have to walk the same distance. The headman, who is a contractor and with whom the order to build it was placed, complied with their request against the wishes of several Anavils.

At the supralocal level, the growing emancipation of the Dhodias begins to have effect. In accordance with the government policy of giving the population more control of their own affairs, *taluka* councils were established in 1963. As chairman in this subdistrict a Dhodia was elected, one of the very few *adivasis* to be elected to such a post in Gujarat. Although he does belong to the Congress Party, locally the instrument of power of the Anavil Brahmans, and was put up as their candidate so as to ensure the support of the village headmen from the eastern part of the *taluka*, the Anavils in Chikhligam see his appointment as the writing on the wall. They are less than happy with present developments, which they regard as threats to their position. Despite the lack of unity among them, they usually take a common stand where their interests are concerned. They are particularly alarmed at the comparatively rapid population increase among the *adivasis*, and blame the government for supporting these tribal castes. In another dozen years, they fear, the Dhodias will have their own political leaders, who will manage their affairs at higher levels than that of the village. Differences of opinion among the Anavils are put aside for the sake of unity. Only mutual cooperation, they argue, will enable them to withstand the much more numerous Dhodias. As a prognosis for the future this may be dubious, but at present the dominance of the Anavils is still undisputed.

7.4 The Pattern of Agricultural Activities

Agriculture in Chikhligam largely depends on the rain, which falls only during the southwest monsoon, from the end of June until the beginning of October. During this period the *taluka* has an average rainfall of about seventy inches, with a possible variation of 50 percent either way. After the wet season, winter sets in, and lasts from the middle of October to the beginning of March, with an average temperature of 60° F. The third season is summer, the hot period in which daytime temperatures rise to over 100° F.

Of the total arable land in Chikhligam (about 800 acres), half consists of the rather infertile and stony *jarayat*. In other parts of south Gujarat, where this soil is more fertile and level, *jowar* (sorghum), cotton, or groundnuts are grown in the monsoon period. My informants affirmed that these crops grew in Chikhligam on the most level high village lands before the Second World War, together with some inferior varieties of grain (*nagli* and *kodra*) and castor seed, from which oil was pressed for use in house lighting. In the agricultural economy of Chikhligam these products probably have never been of great importance. As in other villages on this infertile *jarayat*, their cultivation was abandoned during the interwar depression.

At present the *jarayat* is used only for growing grass, which is cut some weeks after the end of the monsoon. The hay is intended for the owners' own cattle and is not sold outside the village. During the winter and summer months only the *babul* tree, so characteristic of the landscape in this region, grows on the dry and barren soil. Its wood is used as firewood, and its thorny branches serve to enclose the *kyara* and *bagayat* lots as protection against cattle and other wandering animals.

In the *kyari*, which lies scattered among the *jarayat*, only rice is grown. Formerly sugar cane was cultivated here, but this ended several decades ago. At the first rainfall the soil is plowed under and the rice plants are set out in seedbeds. Fertilization is often confined to these small pieces of land. About three weeks later, the topsoil is kneaded with water into mud, and the seedlings are transplanted. No thorough weeding is done. Immediately after the rice harvest, the plots that are still wet enough to plow are sown to *val* (a leguminous plant) as a winter crop. If, because of lack of rain, the rice is late in ripening, the soil is too hard to plow, and in many cases no *val* can be grown. The *kyari* crops are consumed mainly by the owners, and only a few small cultivators sell part of their produce either to other inhabitants of Chikhligam or to merchants in the *taluka* town, usually not because they have a surplus but because they need money. Later in the year they have to buy rice again at a higher price.

Only on the fertile *bagayat*, which constitutes less than 10 percent of the total arable land, is there some well irrigation. At present this garden land is completely under mango. Orchard cultivation, too, depends on rainfall, but in the dry winter and summer months irrigation is desirable. Formerly, twenty-three wells were used in the gardens, water being brought up from them by the labor of yokes of oxen. Only a limited area could be irrigated, and a great many laborers had to be employed as well. At present, four wells are in use, and for these, motor pumps have been installed. Their total capacity is such that all the gardens can be irrigated and a great deal of labor is saved.

The mangoes are picked from the middle of May to the middle of June. During the rest of the year the trees demand little attention. In winter, until the hot season begins in March, the leaves are regularly sprayed with insecticide, and in the dry period occasional irrigation is necessary. Especially when fertilizer has been added, the soil must be plowed; this is usually done at the end of the wet season and in summer. Some large landowners in Chikhligam grow mango seedlings, which they sell even outside the *taluka.* The cultivation of these young trees requires great expertise and care. Kernels of mango fruits are set in rows before the beginning of the monsoon. At the end of the rainy season of the same or the following year, the young trees are potted and put into pits. Daily watering is necessary until the rainy season starts, and the seedlings are then ready to be sold. They are watered every week until April, when summer has begun, and at that time the seedlings are grafted upon mango trees in order to produce the desired species.

The gardens have not always been under this monoculture of mangoes. The first cultivated trees were planted in Chikhligam only forty years ago. Prior to that time, crops in wide variety were cultivated, largely for home consumption, on the *bagayat*: sugar cane, elephant's foot (*suran*), sweet potatoes and other vegetables, bananas, and turmeric. Sugar cane was the only real money crop with which tax was paid, and with which large purchases — of cattle, for example — outside the village were financed.

Reduction of the range of crops has taken place not only in Chikhligam. In the course of a few decades, mango cultivation has superseded the traditional garden culture in the whole of south Gujarat, and the mango has replaced sugar cane as the most important commercial crop. It was especially during the nineteen-forties that mango-growing assumed large-scale proportions. Whereas in south Gujarat the mango orchards covered 4,095 acres in 1938–39, they had increased to 21,538 acres in 1953–54.[6] The owners of *bagayat*, who now depend on the produce of the orchards for their money income from agriculture, devote much more attention to them than to rice-growing. Special care is given to mango trees, as is indicated by the use of manure and other fertilizer,

6. B. A. Desai, 1–10.

thorough weeding, spraying with insecticides, and other measures. This greater concern is also apparent in the cultivation of mango plants, which requires expert knowledge. Fruit-growing has proved so profitable that recently tracts of moderately rough *jarayat*, insofar as they adjoined the *bagayat* area, were leveled and planted to mango trees. After six years the trees begin to bear fruit. So long as they cast little shadow, the traditional *bagayat* crops can be grown among them. Although this was done during the introduction and extension of mango-growing in the existing gardens, the new *jarayat* lots are now confined solely to young mango plants.

The government is attempting to introduce the so-called Japanese method of rice-growing, which guarantees the cultivator a higher yield but requires more labor and capital. The farmer has to level the *kyari*, use superior seed, sow in rows and on elevated beds, observe certain distances between plants, weed intensively, and use more fertilizer. Not a single cultivator in Chikhligam had adopted this method.

As indicated by this brief survey of present and past agricultural economy in Chikhligam, the present crop system is not characterized by highly intensive labor. More than half the arable soil is not cultivated, and agricultural activities are unevenly distributed over the year. They are concentrated mainly in the periods just before and after the rainy season, i.e., the months of April, May, and June, especially in the gardens, and October and November, especially on the *kyari* and *jarayat*.

Inasmuch as agriculture in the village is so extensive, it is impossible for most Dublas — and their number has rapidly increased in the past few decades — to subsist on it the year round. Their seasonal migration to the brickyards fits well into the agrarian cycle; most of the activities fall in the period directly before and after the wet season, when the laborers are still in the village. Nor are their employers, the Anavil Brahmans, much interested in agriculture; they say the profits are too low. More and more Anavils, especially the young, find employment in urban centers.

In the circumstances, the fact that both landlords and agricultural laborers prefer to work outside the village is understandable. To say that it offers them a way out of the restricted possibilities is to state the present position, not to explain it. On inquiry I found that in the past agriculture figured much more largely in the lives of both Anavils and Dublas. How has it come to pass that a large

TABLE 7.4.1
AGRICULTURAL CALENDAR IN CHIKHLIGAM

Month	*Jarayat (grass)*	*Kyari (rice and val)*	*Bagayat (mangoes)*
July through September	After the first rainfall, cutting of green grass	Mixing the soil, transplanting rice seedlings, cutting green grass on embankments, weeding	Cutting green grass, potting young plants
October through November	Making hay some weeks after the end of the monsoon	Reaping and threshing rice, plowing, sowing *val*	Mowing grass in orchards, plowing the ground
December through March	—	Picking *val*	Irrigating orchards, spraying trees with insecticide, watering young potplants
April through June	Cutting down *babul* trees if necessary	Repairing field embankments and fences, plowing ground and fertilizing seed-beds, setting rice seedlings at first rainfall	Grafting young plants on trees and daily watering, fertilizing, irrigating and plowing the ground, pruning trees, mango-picking

proportion of the caste of agricultural laborers now migrates for a considerable period every year, and that more and more members of the dominant caste even leave the village definitively? The answer must be sought in the character of the changes that have taken place during the last few decades.

7.5 The Hali *System in Chikhligam, and Its Disappearance*

"Formerly the Dublas were our devoted servants, they listened to what we said, were satisfied with what we gave them, and worked hard for us. Now everything has changed. The agricultural laborers are in the village only during the rainy season. They have abandoned us and no longer honor their superiors." This was the purport of what some old Anavils told me on the day of my arrival in

Chikhligam. Seated on their porch swings in front of their houses, they reminisced about the good old days and regretted their disappearance. With unending patience and many digressions, members of this caste were prepared to embroider on the theme that day and throughout my stay, for they had nothing else to do.

To inquire after past conditions is risky business. Even information about events of a few years earlier sometimes proved to be unreliable. To reconstruct the relationship that existed several decades ago between landlords and agricultural laborers is more difficult still, in part because young informants have to base their statements on hearsay. I tried to verify the data I was given on past events as carefully as possible, for instance by asking not only Anavils and Dublas but some members of other castes as well about these matters. Comparison of the accounts of agricultural laborers with those of landlords showed that the latter tended to exaggerate by overstating the number of *halis* employed by their fathers. As the discrepancies in the numbers stated by the two groups were considerable I could not quantify the data on this past period. What was important, however, was that all those concerned described the same type of relationship: they all dwelt at length on the *hali* system. The information given me by the villagers on this subject agrees with the picture that emerges from the government reports of the nineteenth and the beginning of the twentieth century which I cited earlier.

The relationship between *hali* and *dhaniamo* was rarely severed. One instance in which it did end during the participants' lifetime involved an Anavil family which is now the most prominent one in the village. At the beginning of this century the family's fortunes were at their lowest ebb. A contemporary official report describes the period as one of famine in the *taluka*, and *halis* were dismissed everywhere.[7] Moreover, three very aged brothers in this family told me their parents' house was destroyed by fire and a large part of the livestock was lost. Their father was obliged to dismiss all his *halis*, twenty of them as they claimed. Aside from such setbacks as this, however, the tie between the families of landlords and agricultural laborers remained intact for many generations.

Why did the Anavil Brahmans in Chikhligam take *halis* into their service? In explaining it to me, both landlords and farm

7. The same report (1897, 7–8): "*Halis* are no longer maintained as the Anaolas find it hard enough to feed themselves, let alone their *Halis*."

servants were inclined to ascribe it to economic motives. As I noted above, agriculture was much more intensive in the past. In the *bagayat* plots along the river, the landlords grew various garden crops, chiefly for home consumption. Most of these, and especially sugar cane, which was a commercial crop, required highly intensive labor. At that time the *jarayat* was also under cultivation. In winter, when work was scarce, the servants took oxcarts to the forest area east of the *taluka* to cut wood for their masters. The necessity of using their laborers throughout the year made it worthwhile for the landowners to take *halis* into their employ. Moreover, the Anavil Brahmans desired to refrain from agricultural labor as much as possible themselves. Not all members of the dominant caste could completely avoid working, but the largest landowners among them did, especially members of the Desai family. Many Bhathela families in Chikhligam, however, still worked with the servants on the land. They now say euphemistically that they occasionally "lent a hand," but they were not in a position even to avoid plowing, a demeaning activity for Brahmans.

The economic situation of the members of the dominant caste determined whether they employed servants, and how many. Most Anavils had one or two *halis*, only the most prominent among them having more. In this respect, too, the situation in Chikhligam conformed with that reported for other villages in the same region at that time. The Anavil Brahmans have always given high priority to their wives' inactivity as well. Where all the men of the caste tried to avoid working on the land, it is obvious that their wives did not assist in the fields. In the house the women could likewise be spared the necessity of doing less favored or defiling work by leaving it to the *hali*'s wife, the maid.

For a Brahman, taking a *hali* was evidence of his prosperity and enhanced the esteem in which he was held. His doing so conformed with what was considered fitting for a member of the caste of landlords. The local Desai family set the tone. As my informants assured me, this family had the greatest number of servants. The Bhathelas in the village tried to emulate this "lordly" example, but mostly in vain. To me they boasted of the many *halis* employed by their fathers, but as I said, on this point their information was unreliable.

To dismiss a *hali* was an action of which the master was ashamed. He did it only in extreme cases, for by dismissing a servant he admitted that he was no longer able to fulfil a patron's obligations of aid and protection. The ill luck of one Anavil in the village was explained to me in one sentence, "His father had five daughters and five *halis*," an allusion to the obligations of this family in the previous generation. To live above one's social position, that is, for a landlord to have more servants than he could afford, was finally to lose land and be compelled to discharge the servants. In this connection I was told about the decline of an Anavil from a neighboring village in the previous generation. As master of a *hali* with six sons, this man had been so conceited as to think it humiliating not to make use of his preferential rights. To avoid what he thought of as loss of face he took all six of them into his employ by giving them marriage money. My informants gave me graphic descriptions of these Dublas doing nothing and living at the expense of their master. Pride once more went before a fall: when the man died, his sons were left without resources. The greater part of the land was lost, and the servants went to another master.

Many Anavils had to work with their servants on the land, and some had no servant at all. Only occasionally did these poorer Anavils use the labor of the unattached Dublas in the village. For them the seignorial life style of the larger landowners remained an unattainable ideal.

Among the Dublas, the first consideration in becoming *halis* was the security they found in servitude. Not all members of this caste of agricultural laborers succeeded in finding a master, an Anavil who was prepared to provide them with a livelihood. Both Anavils and Dublas confirmed that there had always been a category of casual laborers in the village. It might be important to know the former proportions of unattached laborers and *halis* in the village, but the statements I have on this ratio are not exact and diverge widely. Most of the informants of both castes, however, stated that the number of farm servants was many times larger than it is now. They mentioned especially the smaller size of the Dubla caste at that time.

The casual laborers, too, mainly depended on the landlords in Chikhligam for their livelihood, but they added to their earnings by work in the surrounding villages. Anavils and Dublas were unanimous in their belief that before the Second World War the *halis* were better off than the casual laborers. Especially during the slack winter months, the unattached Dublas often had no income and led a miserable existence.

This is the background against which must be judged the accounts of the affection and loyalty felt by the servants for their masters, so highly praised in retrospect by the Anavils of the village. Evidently, to put his security at stake by risking a conflict with his *dhaniamo* was the last thing a *hali* would do. Moreover, such norms as obedience and fidelity were guiding principles in the Dubla mind. Running away was a sin for which a Dubla would be punished in another life by having to serve his master as an ox. Thus the behavior of a "good" servant was supported by religious norms.[8] A more immediately operative sanction was that a deserted master could count on the cooperation of the local authorities, themselves usually caste members, in hunting down the Dubla.

For another Anavil to take on a servant without the permission of his former master was also prohibited. Things were different when, about thirty years ago, a Mochi bought *bagayat* land in Chikhligam with money he had earned in East Africa and took a Dubla into his employ by giving him marriage money just as the Anavil landlords did. His venture failed, as the Anavils took pleasure in telling me. This Mochi *dhaniamo*, of low caste and incompetent as a landlord, did not compel respect. His servant's impudence and laziness were too much for him. After a time the Dubla entered the service of an Anavil. The new master, however, did not pay the Mochi for the debt incurred by this *hali*, as would have been usual. In other ways, too, the Mochi, the only known example in the village of a *dhaniamo* of low caste, came to grief. He leased out his land on a crop-sharing basis to an Anavil landlord, and when the latest tenancy reform occurred this Anavil had himself registered as the owner.

On the other hand one should not take an exaggerated view of the collaboration and harmony within the dominant caste in the past. The relationship between the various patrons, who were divided into factions, was one of competition. By luring a servant away from a fellow caste member they could cause the latter to lose prestige, for this proved that his servant was not loyal. Here was a

8. "The Dhaniamas, belonging to the higher castes, have sedulously developed a deep-rooted feeling among the Halis that running away from the master is a great sin, for which God will punish the Hali's descendants." In *Report of the Congress Agrarian Reforms Committee*, 1951, 129. See also Lorenzo, 1943, 66.

possibility, although a very small one, for the Dublas to manipulate their masters by reminding them of their obligations as patrons and addressing them as such in the presence of others. Nevertheless, the pressure which the *dhaniamo* could apply was considerable, and the chances for his subordinates to escape it were slight. "They did anything we asked of them or more; they were our slaves (*gulam*)," the Anavils of Chikhligam now say.

In a situation of complete powerlessness the only thing to do is to recognize one's dependence. By clearly showing subservience and docility the servant could try to gain extra advantages and to receive better treatment than his fellow servants, thus persuading himself that his position was higher than theirs. A Dubla prided himself on belonging to the household of a prominent Anavil. This attitude has not yet wholly disappeared. An old Dubla, the servant of a Desai in the village, whispered to me that his master was the only true Anavil in the village, the other landlords being only Bhathelas.

Despite the inferior position in which the members of the castes of agricultural laborers found themselves, servitude offered them opportunity to gain some influence and esteem, if only among their equals. Tales about the good old times, the bond and mutual sympathy that used to exist between master and servant, are told even now by the older Anavils and Dublas. Concerning the traditional situation in Tanjore, Gough writes:

> In recalling the past, it is the mutual love between master and servant, not the Brahmans' power to make unlimited exactions, in which they take pride. Powerful economic, ritual and emotional ties bound each Brahman household to its own hereditary labourers.[9]

The statements of landlords and agricultural laborers in Chikhligam alike confirm the conclusion drawn in chapter 4 about the situation that obtained some decades ago: they were partners in a system of bondage and patronage.

The lines on which the relationship between landlords and farm servants in Chikhligam are now organized are quite different. The change is recent and is closely bound up with the introduction of mango cultivation just before the Second World War. The mango was not a new fruit to the population of Chikhligam: they knew it as a country fruit for their own use, which was picked from

9. Gough, 1960, 50. Similar statements can be found in several other anthropological publications on village India.

trees growing on the enclosures or in their yards. The first man to cultivate mangoes was the late husband of an Anavil woman who now manages the family property for her sons who work in town. All fellow caste members have followed his example. It was a small-scale change at first, and in the initial stage other crops were still grown among the young trees. This interesting possibility of lowering the cost of investment has been wholly neglected in the recent extension of the orchards to favorably situated *jarayat* lots. When I asked why, the Anavils replied that it was too much trouble.

The changeover to the relatively nonintensive cultivation of fruit trees cannot be explained by an existing or imminent shortage of labor. Although, as I was told, some Dublas worked in the brick-yards near Bombay for a few months of the year, their number was small. Such employment as existed outside agriculture before the war became even scarcer during the depression years. Moreover, my informants emphasized, the Dublas were practically all attached at the time. The landlords would never have permitted their laborers to leave the village in large numbers. Only when there was less to do in agriculture, when, in other words, the Anavil Brahmans no longer needed them, did the yearly migration to the brickyards get into its swing. Eight older Dublas told me that they had begun work as *halis*, but that they had been dismissed as redundant. This was confirmed by their former masters. It was not the increase of alternative employment in town that was responsible for the seasonal migration of the members of the caste of agricultural laborers, but rather their redundancy in the village economy.

Anavil landlords themselves mentioned the following reasons for the change of crops: plant diseases in some garden crops, the fall of cane-sugar prices, and the growing demand for mangoes in the towns. Although such considerations in the field of agronomy and market economy no doubt contributed to the replacement of the traditional *bagayat* crops, they do not explain the preference for mango cultivation in the localities inhabited by Anavil Brahmans. The role played by social factors in the introduction and diffusion of the new crop was, in my view, more important. From the beginning the members of the dominant caste had strived after a seignorial status, but their active participation in agriculture was unavoidable even if they did not till the land themselves. They had to be continually present in their fields or gardens to supervise and direct the work of their farm servants. Fruit-growing, however, enabled them to expend even less energy on managing their property. Especially for the Bhathelas among them, who could not afford much help and of necessity worked their own land, the cultivation of mangoes meant almost complete deliverance from personal activity. A changeover to this crop requiring little labor enabled them to lead the same kind of life as their more well-to-do fellow caste members. V. H. Joshi has correctly pointed out that the introduction of mango cultivation in south Gujarat was to some extent due to the Anavils' desire for a higher money income, therefore to a commercialization of agriculture.[10] It is, however, doubtful whether the landlords gave evidence of a capitalist mentality — in the sense of careful calculation of marketing possibilities for their agrarian output — when they so clearly were motivated by their wish to exert themselves as little as possible. An illuminating comment by G. Keatinge on the rapid expansion of cotton cultivation in central Gujarat in the first decades of the twentieth century points to the same conclusion:

> It is even asserted by careful observers that the keenness of cultivators to grow cotton is due not only to the fact that they can usually make good profits from it, but also that it is an easy crop to grow and leaves them plenty of leisure.[11]

So far the change in the crop system seems to fit in completely with the long-standing ideal of the members of the dominant caste to live in lordly style. At the same time, their wish for a less active role in managing their property also arose from a more general loss of interest in agriculture and in village life. Even in the previous generation, the Anavil Brahmans showed an increasing preference for living and working outside the village. This tendency has increased among the present generation, as I shall substantiate below. It is obvious that the lessening of the landlords' involvement in agriculture and village life has had far-reaching consequences for the position of the Dubla agricultural laborers.

Thus, the relationship of patronage between landlords and agricultural laborers ended. The time was past when the Anavils felt obliged to provide for Dublas by taking them into their employ as *halis* and thereby recognizing them as clients who were entitled to aid and protection. That the landlords were no longer able to take on a large proportion of the agricultural laborers, owing to the

10. Joshi, 1966, 49.
11. Keatinge, 1921, 146.

demographic shift between the two categories, was also a factor. In recent years the number of Dublas in Chikhligam has risen rapidly. I was told by various informants that there are now many more huts of this caste than there used to be. The number of Anavils, on the other hand, has decreased. As noted above, a relatively large number of them have left the village permanently, while the natural increase of the caste in the village has remained below that of other castes. The result has been that the ratio of agricultural laborers to Anavil households has changed. For the latter it was simply impossible to provide the same proportion of Dublas with a permanent living and, in general, to behave like patrons towards them. Nor were they sufficiently motivated to endure any privations in order to do so.

I have not found any indications that the disappearance of the *hali* system in Chikhligam could be ascribed to outside pressure. The British authorities in south Gujarat did not in any way intervene in favor of the agricultural laborers, and "social workers" who, influenced by the Gandhian movement, fought for the interests of the Dublas were not noticeably active in this *taluka.*[12] At most it can be said that some landlords were vulnerable to the criticism, especially emanating from Gandhian circles, that they practiced complete bondage of agricultural laborers in the period of *halipratha.* At the same time, however, they argue inconsistently that the greater freedom of the Dublas has simply led to more ingratitude. It annoys them that complaints about the past, however justified, are extended to the present.

A changing orientation of the landlords was, I think, of decisive importance, and the changeover to mango cultivation was an expression of it. The wide range of obligations to *halis* and their families was increasingly felt to be a burden. By means of detailed calculations, Joshi shows that in the span of a year a *hali* proved to be much more expensive than a casual laborer who was needed for only a limited number of days.[13] These views concur with those in some earlier reports in which the *hali* system was called uneconomic, a statement that was usually elucidated by a comparison of cost as between a day-laborer and a farm servant, which was always to the detriment of the latter.[14] It remains to be seen whether the

12. *Report of the Hali Labour Enquiry Committee*, 52.
13. Joshi, *op. cit.*, 56–57.
14. *Census of India*, 1921, vol. III, pt. 1, 222; Mukhtyar, *op. cit.*, 162–66; Shukla, *op. cit.*, 121–25.

return of a *hali*'s labor in the past was relatively much higher. At any rate, those who put so much stress on wage cost and labor productivity do not distinguish the element of patronage inherent in the *hali* system. Although the population of agricultural laborers has indeed proportionally increased, the "surplus of labor" is especially the result of a different definition of the laborer's economic use, this time solely in terms of work performance.

In their dealings with Dublas, the Anavils generally assumed the attitude of employers toward employees. The question they put to themselves was no longer "How many farm servants can I provide for?" but "How many hands do I need, and when?"

It was from sheer necessity that the Dublas, no longer assured of subsistence, reacted by partially withdrawing from the village economy, which they could do because of the opportunity of seasonal employment in the brickyards near Bombay. In my discussion of the conditions I found in the course of my field work I shall elucidate this further.

7.6 The Present Situation

THE DUBLAS. My data on the Dublas relate to fifty-one huts with 327 occupants in all, that is, an average of 6.4 per hut. The village nucleus comprises twenty-six huts, and twenty-five have been built just outside the territory of Chikhligam. Of the male working population over sixteen years of age — ninety-five by my count — thirty-eight are employed in agriculture. Most of them work as farm servants (twenty-one) or as day-laborers (eleven), the others (six) cultivate their own land and add to their meager income by regularly performing field labor for the landlords of Chikhligam (see subsection 7.2). Five Dublas are employed in the village but outside agriculture, and at the time of my stay, fifty-two had gone to the brickyards.

When a man marries, a new household is usually formed and new accommodation is sought. Of the fifty-one huts, thirty-three contain a household of two generations, usually parents (or a parent) with unmarried children. The joint-family character of other households is found chiefly among those Dublas in the village who are permanently employed outside agriculture or are landowners or both (eleven in all). Compared with their landless fellow caste members, these Dublas are economically better off also because their position is augmented by a greater number of male earners per hut (averaging 2.8 as against 1.3).

In Chikhligam, as in other villages of south Gujarat where the gardens have been transformed into orchards within the past few decades, employment in agriculture has declined sharply. Attached labor, however, has not declined. A surprisingly large number of Dublas (twenty-one) are still farm servants; one is employed by the Bania of the village, the others by Anavil landlords. More than half of them work for the most active members of the dominant caste in the village — three Vashi families — who not only cultivate fruit trees but also grow mango plants, which require more, and more frequently applied, labor. Most of the other Anavil households also have one or more servants. They themselves do not work, and the size of their property is such that it is easier for them, despite the relatively nonintensive crop cultivation, to appoint a Dubla on a yearly basis than to employ day labor regularly at small intervals. The oldest farm servants in the village entered service as *halis* thirty or more years ago, that is they married at the expense of their fathers' masters after having worked for them as cowherds. They are exceptional, however, and most farm servants today entered service at a later age, and only recently. When younger, nearly all of them worked in the brickyards for a number of years. Some found the work too heavy, but others say they will certainly go again sometime in the near future. In contrast to its former state, the service relationship is now a temporary one. The servant status is no longer more or less automatically transferred from father to son. Neither party deliberately intends to maintain the relation for life. Landlords as well as farm servants reckon with the possibility — indeed, the probability — that their agreement will come to an end. The Dubla, who is no longer of the sort who worked for the master from childhood, knows that he can find work in the brickyards if necessary, and he has become less compliant. The Anavil aims to give his servant as little credit as possible, so as to limit possible monetary loss in advance. Neither puts trust in the other's intentions, partly because the personal tie between them is lost. Servitude, in fact, has been reduced to a businesslike agreement with a limited purpose between partners who mutually distrust each other.

This is also evident from the Dubla's work performance. At the peak of the season he works from 5 or 6 a.m. to 7 p.m. At other times he begins a little later and stops earlier, and his midday break is longer. Jobs in and around the master's house are not much to his liking. The servant no longer wishes to be at his landlord's disposal from morning till night, and tries to limit his obligations to agricultural labor if he can. Nor does he take it for granted that his wife should work as a maid in the master's household, and his son as a cowherd. The Dubla women working in Anavil houses in the morning are by no means all married to servants. *Govalia*, too, are now separately hired. The contract, therefore, is not only temporary but restricted, relating only to the labor of the servant and no longer including that of the members of his household as a matter of course. Nevertheless, the close relatives of the Dublas are still expected to hold themselves at the master's disposal as extra hands at harvest time.

The remuneration of Dublas in permanent service has likewise become more specific. In kind, the daily ration has remained at a level that has been fixed since the last century: two *seers* (a cubic measure, weighed out in a hollow wooden cylinder) of *jowar* or four *seers* of unhusked rice.[15] In money, their daily wage at the time of my field work was 8 annas. On the mornings when he begins early, the servant is entitled to a sort of breakfast (a *rotli* and some vegetables) and a cup of tea. For some activities he also receives a midday meal. In addition there are the time-honored perquisites: some garments, a pair of shoes, a shawl (also used as a blanket), reeds for roof-mending, permission to gather wood in the gardens, a few sheaves of rice or some *val* at harvest, *chas* (the skimmed milk left over from the preparation of *ghi*), etc. The Anavils manipulate these wage elements as means of pressure to make an unwilling and misbehaving servant see reason.

Some landlords attempt in every possible way to skimp the extras given the servants, and increasingly they prefer to replace the emoluments by money. For the maid and the cowherd this has become the rule. Although the fourteen Dubla women who work in the houses of the Anavils in the mornings still receive a midday meal (rice and *dal*) when they have to stay later than noon, their money wages are about 1 rupee a month. The cowherds, too (eleven in all), who earn 1 or 2 rupees a month, receive

15. Mukhtyar has calculated that on a daily ration of five *seers* of unhusked rice a Dubla family of five persons suffers from chronic undernourishment (Mukhtyar, appendix VIII).

no clothing or shoes, only their food. The Anavil women consider it inconvenient and beneath their dignity to prepare food for Dubla farm servants. The changeover to money payment is undoubtedly due in part to the rise of different habits of consumption. One Anavil, who wished to adopt the kind of agreement prevailing in another village, offered to pay his servants a daily wage of 1 rupee and stop all allowances in kind. Understandably enough, they did not accept his proposal. Transformation of allowances into money payments is to the servants' disadvantage. In the past, a *hali* was allowed to cultivate a plot himself, and the produce it yielded was an important supplement to his daily earnings. During the war years, a period of high grain prices, this *vavla* was replaced by a yearly payment, and in Chikhligam this bonus now amounts to between 25 and 50 rupees, depending on the skill and bargaining position of the farm servant. It is paid in small amounts spread over the whole year, and only when the servant asks for it and the master has no objection. Although originally intended to cover unforeseen expenses, the wage is now necessarily spent on articles that the servant used to receive from his master, on his daily ration of tobacco and oil (formerly pressed from the castor seeds he cultivated on the *jarayat*), and on clothing, shoes, and the like.

Around *divali*, when the accounts are settled, the servant rarely finds that there is money owing to him. Not without reason, the Dublas suspect their master of double-crossing them. On the other hand, the master is likely to deduct from the year's allowance the drink money he has occasionally handed over to persuade the servant to do something outside his routine activities, or else he writes it down as advance payment if the amount is exceeded. This seems reasonable to the Anavils. Formerly the Dubla was at his master's beck and call every hour of the day without receiving extra pay for such service.

The allowance received by the servant for days when he does not work — because of heavy rain in the monsoon period — is also debited to him. Although the master is expected in any case to give this *khavati* (literally, "in order to eat"), he deducts it from the year's allowance. The debt run up in this way, sometimes added to by marriage money, also becomes collectable at *divali*. The Anavils admit that a Dubla cannot possibily discharge the debt so long as he works as a servant, but when he runs away to the brickyards, as farm servants frequently do, the Anavil tries to recover the debt from him there or, as I witnessed, even holds his relatives responsible if they have remained in the village.

This increasingly businesslike character has also entered the relationships among the Anavils themselves. When a Dubla farm servant finds employment with a new master, the latter no longer considers himself answerable, as would have been true in the past, for the obligations incurred by his subordinate toward a previous employer. Even within his own village there is no longer any question of compensation on the part of the new employer. In contrast to former usage, the servant himself is now held responsible.

Conditions of employment for Dublas in permanent service display much that is reminiscent of the *hali* system, but some differences are fundamental. The relationship is no longer hereditary and remains essentially confined to a work performance by one party and an exactly calculated remuneration by the other. The Dubla does not feel bound to do more than his agreed portion of the work, and as for the Anavil, he does not consider himself obliged to pay more than the minimal amount of compensation. If the servant says he cannot live on it, that is regarded as his own problem. There is no longer any question of complete and lifelong support. Two servants in the village who are too old to work are still to some extent useful as *govalia.* But three others, who had worked for their master from childhood, were dismissed a year ago, in certain instances after a provoked quarrel, and are now, sparingly, supported by their children. Nor is continuance of the grain allowance during illness a self-evident right. Although the Dubla may leave his landlord more easily, it will be clear that it has been he who came off the worse when their relationship changed. This is largely the result of the increase of monetary transactions in an agrarian subsistence economy.

Aside from the Dublas in the permanent service of landlords, there is a category of casual laborers. When classifying the economic activities of Dubla men in the productive age group in Chikhligam, I arrived at the figure of eleven day-laborers. They, too, chiefly depend on the Anavils for their livelihood. It is true that casual laborers are free to work outside the village, but only some Kolis in the neighboring village employ them occasionally. These Kolis, like the landowners among the other castes in Chikhligam, are petty cultivators who can usually manage the work with their own families. Inasmuch as the Anavils do not continually need the services of these laborers, there is employment for them during only part of the year. The advantage of a slightly higher wage — they received 12 annas a day during my stay in the village — is outweighed by frequent long periods of unemployment. Their freedom is therefore rather debatable. On days without work they ask the large Anavil landowners of the village for *khavati*, and in compensation they bind themselves to work for the landowners on days selected by the latter. Those Anavils who can afford it make sure of a labor reserve at the peak of the season by giving advances in this manner. Moreover, they pay casual laborers only 8 annas a day during these periods, the same wage as the farm servants. The day-laborers much prefer to go to the Kolis who, they say, make them work harder but are more liberal and join in working with them in the field. They give the laborers 1 rupee and also spirits. These small farmers of a lower caste are, in short, less high-handed than the Anavil Brahmans.

The prospect of a day-laborer's life is not very inviting, and most Dublas in this category regard it as a temporary solution. Not, for instance, the three who were dismissed in old age and now try to earn a little here and there as day-laborers, but those who are prevented by circumstances (a pregnant wife, a baby newly born, or aged parents to be cared for) from going to the brick-yards. Work in the brickyard, not permanent service, is for them the alternative to the meager existence of the day-laborer. Their stay in Chikhligam is a breathing space; those who are young and healthy hope to go away again in the following year.

Considering all this, it is not surprising that the majority of Dublas is on the move every year. There is virtually no hut from which one or more occupants are not absent, working in the brick-yards. Of the men in the productive age-group, fifty-two were absent during the period of my field work. In most cases the whole family goes, so as to add the earning power of the woman and the children. Only the aged, women with very young children, and otherwise unproductive family members remain behind. They usually must provide for themselves, as weekly allowances in the brickyards are low and strictly for those who work there, so that little or nothing is left to support those at home.

The organization of this migratory labor has already been explained above. Seasonal employment in the brickyards has undoubtedly lessened the dependence of the Dublas on the landlords in the village. Assisted by relatives, some manage to save enough for their marriage money. This is exceptional, however, for once in the brickyards the Dublas adopt an easier and less frugal way of life. They eat more and better food — for instance, the meat and fish they cannot afford in the village — take tea with milk, and buy various cheap commodities ranging from clothes to household utensils. Moreover, a great deal of drinking goes on. The amount they finally save is therefore hardly ever enough to pay the costs of a marriage. Occasionally the labor-recruiter or the owner of the brickyard is prepared to help. A seasonal laborer who is known as diligent, honest, and well-behaved can obtain an advance if he promises to hold himself at the disposal of his creditor until the debt is paid off. In this way there results a relationship of dependence of the same type as that with which the Dubla was familiar in the village community.[16]

During the past decade, labor requirements in the brickyards of Bombay have risen sharply. This industry, however, is very susceptible to economic fluctuations and a regression will undoubtedly lead to large-scale shut-downs. Even in the present period of expansion, the migratory laborer's existence is precarious and uncertain. Most of the brickyards are small enterprises with no more than two or three kilns, badly managed, scantily financed, and operating in an unstable market. Marginal production frequently leads to an owner's bankruptcy. so that the laborers have to return to the village prematurely and without earnings. The labor-recruiter is another uncertain element. For placement, the gangs recruited by him depend on the contacts that he has established, often over a period of many years. If he drops out, or if he quarrels with the brickyard owners, it is the laborers who suffer for it. Thus the Dublas are not at all sure of finding work every year, whether in the same brickyard or in another. One Anavil told me that he was convinced that the majority of the migratory laborers would prefer to work for him if he guaranteed them a daily wage of a little over 1 rupee all the year round. But in my view he underestimates the strong desire that has arisen among the Dublas to escape, if only temporarily, from the yoke represented by landlords and by agricultural work. On the other hand, the Anavil's claim confirms my impression that seasonal migration does not hold any great advantages for the Dublas.

16. For a similar type of relationship among migratory labor in Japan, cf. Bennett and Ishino, 1963, ch. 8.

The temporary or permanent absence of Dubla men from the village may lead to termination of marriages. The women who remain behind without support return to their native villages with their children, or they marry again. Divorce occurs frequently (I counted sixteen cases), for instance, when one of the partners is in poor health and cannot work. Many children grow up in the households of relatives owing to the absence of one or both parents; in Chikhligam twenty-three children have been taken into households to which neither of their parents belongs. Similarly, a number of incomplete families have joined other households: I found nine cases of Dubla men or women who had been left behind and who, with their children, had moved in with their parents, a brother or a sister. The relatively low stability of Dubla households — which, I regret to say, struck me as significant only when I was writing up my data — is undoubtedly connected with the precariousness of their existence.

Only a small number of the Dublas from Chikhligam, nineteen in all, have succeeded in finding regular work in town. Nonagrarian employment is scarce and not very accessible to members of a caste of unskilled agricultural laborers unfamiliar with the industrial scene. Nor are there as yet many Dublas who are established in town and can help their fellow caste members in finding work and living quarters. Seasonal migration as a first step towards permanent settlement in town is therefore a solution available to no more than a few Dubas, and most of them return year after year to the village at the end of the dry spell. As they grow older, they usually are completely reabsorbed into the village economy. Work in the brickyards is arduous, the largest contingent there consists of young people, with their greater endurance. The farm hands in the village are mostly older men, who feel no longer capable of the exertions demanded by brickyard employment.

For most Dublas, the escape from agricultural labor by seasonal migration is temporary in yet another respect. The amount of money they bring home from the brickyard — varying with the size of the family, but usually not exceeding 50 to 75 rupees — is not sufficient to tide them over the rainy season. Even with advances borrowed from the labor-recruiter they are rarely able to make ends meet. During the months when they are in Chikhligam, most Dublas are therefore compelled to work in agriculture. At the time of the inquiry, about half of them — twenty-five of the fifty-two Dublas who had been away — had put themselves at a landlord's continuous disposal during the wet season. This category I have described as monsoon servants. Their tasks consist of cutting a head-load of green grass daily and of weeding occasionally. At the beginning and end of the rainy season, with much work to be done in the fields, activities multiply (see table 7.4.1). The workers receive the daily wages of servants on a yearly basis, but not the perquisites to which the latter are entitled. Unlike the Dublas in permanent service, the monsoon servants are not entitled to *khavati* on days when there is little or nothing for them to do, although they try to wheedle it out of their employers. As for the Anavil, he is always assured of cheap labor in this way.

One may ask what induces a Dubla to bind himself under such unrewarding conditions. Frequently an existing dependence on the landlord leaves him no choice. The "obligations" of a monsoon servant may be purely personal. In several cases, Dubla migratory laborers had been employed originally as farm servants, but they had run away to the brickyards without discharging their debts. Seemingly, the members of the dominant caste are resigned to this, for them, disagreeable habit, but on the laborer's return from Bombay they demand payment, and sometimes even before that.[17] So long as the Dubla is indebted, and this may be permanently, he is obliged to work for his former master in the rainy season. In other cases the pressure is more direct. The landlord sees to it that on their return he will be the only one to make use of the services of relatives and certainly of members of the household of his permanent servant. After a few days — to avoid giving an impression of wanting them badly — the Anavil goes to the hut in which they live, asks after their health, and tells them when he expects them. For members of the dominant caste, the possibility of obtaining an extra amount of labor and, in this way, protecting themselves in advance against shortage, is an important consideration in favor of taking a Dubla into permanent service during the rainy season. At that period there are usually enough agricultural laborers available, but at some critical moments the demand for them is so high that even if the supply is maximal, i.e., if all the Dublas are there, not all the landowners can procure the labor they need. In the competition for laborers it is the largest landowners who come out best, as I noticed at the beginning of the rice harvest. Through

17. By visiting a number of brickyards in Bombay I tried to gain an impression of the migratory labor. On one occasion I observed that an Anavil came and fetched a runaway servant home.

various "obligations," some landlords had been able to monopolize the labor of a large part of the Dubla work-force, and moreover, against low wages. While the Dublas worked with their families in the fields of their "benefactors" as servants, monsoon servants, and day-laborers entitled to *khavati*, the smaller landowners had to delay harvesting until the others had finished, pay higher wages to the Dublas then available, and in some cases even give up sowing the winter crop, as by that time their soil had become too dry to be plowed.

Seasonal migration has become a vital necessity for the majority of Dublas. Though such migration is partly due to the rapid growth of their numbers, the changeover to mango cultivation is chiefly responsible for the fact that so few of them find continuous employment in agriculture.

The yearly migration to the brickyards fits neatly into the agricultural calendar. The Dublas' presence in Chikhligam coincides with the period of greatest activity on the land. This is not as fortuitous as it seems, for although they could have found work in Bombay at an earlier date, the Dublas who work as monsoon servant are not allowed to leave until after the grass harvest (the middle of November), when the interests of the landlords are safe. The Anavils have discharged their obligations towards the Dublas, i.e., to assure their livelihood, but have managed to retain their traditional claim to the working-power of the members of this caste. Although the seasonal migration has lessened the Dublas' dependence, it has not put an end to it. Many of the Dublas are temporarily bound to work as monsoon servants, but in the long run most of them are reintegrated into the village economy as servants or day-laborers. As one of the Anavils succinctly put it, the Dublas work for them on their conditions, "under obligation of debt and for future credit."

THE ANAVIL BRAHMANS. The seasonal migration of the Dublas from Chikhligam was an effect, not the cause, of a shift towards a less varied and labor-intensive agrarian economy. In the preceding pages, I interpreted the introduction of mango cultivation as primarily a reflection of the lessening involvement of the Anavil Brahmans in agriculture. The cultivation of the traditional garden crops is undoubtedly much more profitable, but it would require of the landowners an amount of exertion and trouble which they are not prepared to devote to it. Already in the past generation several members of the dominant caste went to live in town, and recently this movement has gained impetus. Of the male Anavils in the productive age group, distributed over fourteen households in the village, sixteen live on their land in the village, thirteen are employed in nonagrarian occupations, and all but one in the latter category live outside Chikhligam (see subsection 7.2). Three Anavil landlords have additional incomes, respectively as a contractor, a postal agent, and an insurance representative.

As town amenities increased, village life became less attractive to the dominant caste. The young people in particular yearn to live and work in the town. Urban Anavils are preponderantly professional men, clerks with the government or in the private sector, and teachers in primary and secondary schools. It is the typical white-collar work in the administrative sector that draws the young Anavils. Education has become an imperative necessity for them, as they need an S.S.C. (secondary-school certificate) to qualify for even a simple administrative position. The poorer Anavils of the village also send their children to school and count on the help of relatives and fellow caste members in town to find them jobs. The household is prostrated if the son fails in his examination. Again and again the candidate tries to surmount this barrier to a better social position, mooning about the village between examinations. One landlord was overcome with joy when his unclever daughter, after having failed four times, managed to cross the finish line after all. Even without a high dowry she was now regarded as more eligible to marry an urban fellow caste member. Through schooling, the younger generation tries to escape the "backward" village life, and the older generation helps them.

Those who remain behind in Chikhligam feel that they have had bad luck and look upon themselves as misfits. They have not been able to leave the village, and long for the amenities and pleasures of town life. Without much zest, and under the pressure of circumstances, they manage the family property and complain that agriculture is not very profitable. This argument seems ill-founded, for in the first place it assumes as a norm the most successful town-dwelling Anavils, not the majority of their fellow caste members who lead simple clerks' lives. Moreover, a more active management and more investment in agriculture would easily better their situation. But the village Anavils endeavor as conspicuously as possible to be nonfarmers. Their dilemma is that they want to gain as high an income as they can while working as little as possible. Although they rarely visit the *jarayat* or *kyari* plots they possess, most of them take a walk to their gardens in the afternoons in order to inform themselves of the state of the trees. Wearing white *topis* and carrying impressive long canes, they demonstrate their position of landlords and inspect the work of their laborers.

The cultivation of mangoes is the main topic of conversation. "The king of fruits for the kings among men" was the slogan they repeated to me with a mixture of self-mockery and conceit. Seated on their porch swings and looking out over their front yards, they never tire of discussing the quality of the bloom, prospects of the crop, marketing possibilities, price fluctuations, and such matters.[18]

Although most of the village Anavils have been able to abstain from agricultural labor, two of them cannot afford the luxury of not working. These two do not possess much land and are not so comfortably housed as their fellow caste members. Yet they conform as much as possible to the way of life they consider fitting for an Anavil. For plowing, an activity despised by Brahmans, they hire Dhodias. Day-laborers help them with the rest of the work in the fields. One of them also employs a Dubla woman to do the dirty housework. Their children go to school, and one son is a post office worker in the *taluka* town. His father has managed to offer his children a better future, but it is doubtful whether the new generation of the other Anavil family will ever be in a better position. This family is, unfortunately, very large and, worse still, includes many daughters for whom marriage money must be set aside.[19] Unlike the other members of his caste this Anavil has not had his wife sterilized, and the consequences prevent him from improving his weak economic position. As working cultivators, these two Anavils strike a discordant note in their caste.

Eight Anavils have one or more farm servants in their employ (twenty-one in all). Supervising their Dublas is the only activity they engage in. Some of them lend a hand occasionally, but real work is left to their servants, supplemented by day-laborers if necessary. Most of the masters confine themselves to the shouting of instructions or abuse. Four Anavils do not have enough land to employ a farm

18. Cf. also Joshi, *op. cit.*, 53.

19. As to the amounts of money required for such marriages, see Van der Veen, *op. cit.*, 97–99 and 135–37.

servant, but feel that to work would be beneath their dignity. They have leased out their *kyari* land on a crop-sharing basis to fellow caste members who in this way can fully utilize the working power of their own servants. The produce of their orchards is usually sold in advance (when the trees are in bloom) to a factor outside the village who takes care of all the work involved. Their income might be higher if they exerted themselves, but some supplement it with part of the earnings of a son in town. They have a simple routine: morning tea and breakfast — bath — meal — afternoon nap — walk to the garden — tea — evening meal. In the evening as in the daytime they can be found on their porch swings in front of their houses. They are landlords so far as their life style is concerned, but certainly not farmers.

Those Anavils who employ farm servants are more active, for they supervise the work of the Dublas during mango-picking and the rice and *val* harvests, plan the day's work with their laborers, visit their gardens more frequently than the others and also grow young mango plants. Nevertheless, they are not overly busy. Illustrative of the indolence of this caste is the Anavil who has his servant carry an easy-chair to the orchard, so that he can watch the work of his underlings in comfort.

The reputation of the landlords of today — that they have forgotten how to drive an oxcart or to plow a straight furrow — is a dubious compliment in the eyes of members of other castes. Generally speaking, the Anavils themselves are in no way ashamed of their sometimes feigned ignorance in such respects, but the most competent of them regret that so many of their fellow caste members have thrown all responsibility in agriculture to the winds, even that of overseeing the work performance of their laborers.

Still another instance of their lack of interest in agriculture may be cited. In 1962 Chikhligam was connected up with the irrigation network of the large Kakrapar project in south Gujarat by the extension of a branch canal. At no time during my stay was there any indication that the landlords were going to seize the opportunities thus offered by, for example, leveling and irrigating the *jarayat*. The new potential was at best regarded as a reserve in the event that rainfall at the end of the monsoon should fall below expectations. Nor have Anavils in the surrounding villages, where irrigation has been available for a longer period, taken any steps towards intensification of agriculture.

Mango-growing has, above all, made possible a "more dignified" style of life for all Anavil Brahmans. The distance between members of the lower Bhathela section and the Desais, which must have been considerable, has lessened, and the difference in life style between them has faded. During the previous generation, the Desai family had already invested in education, for which those who remained in the village sold their land. Their former gardens are now largely in the hands of the Vashi family, who are, in their turn, spending their acquired riches on nonagrarian purposes. The young people all receive educations that are meant to have them end up in town. "Conspicuous leisure" has become possible for nearly all Anavils, who — with or without servants — can now lead a landlord's life, or at least pretend to. They no longer derive prestige from the number of their laborers, or in general from the number of clients of other castes retained by them. The dominant caste in Chikhligam is still divided into factions, but for exercising influence at the village level the support of a group of local clients is no longer the only relevant factor. As the economic, political, and administrative scale was enlarged, the display of power within the village community declined and became subject to external structures of decision-making. In this respect, the former patrons have become clients. The success of their activities outside the local framework is partly determined by the weight and esteem they have acquired on other grounds than the criteria of former times. On the other hand, to have a body of support in the village is to be able to exert influence legitimately at a higher level. In this form, patronage has come to have a more restricted purpose and usually remains limited to specifically political matters. Conversely it is true that connections with sources of power — particularly with the government bureaucracy and with the political parties located in town — is still a sovereign means of commanding respect and support in the village. Then, too, in their new scale of desiderata the Anavil Brahmans give the highest priority to education. The profits returned by this investment are considerable, because they enable the Anavils to orient themselves towards town life.

Thus the traditional patron-client relationship has come to an end. Obviously such an important institution as patronage — a whole network of personal relationships resting on very pronounced norms and values — did not disappear overnight. In Chikhligam there are still recognizably patronage-like elements in the behavior of Anavil Brahmans. That the personal character of the landlord concerned makes a difference is illustrated in the attitudes of the village headman and his cousin (the son of his father's brother). They belong to the prominent Vashi family and own an equal amount of land. But whereas three farm servants are enough for the village headman, his cousin requires no less than six. The latter also employs more Dubla women in his household, one of whom is solely charged with the care of the children. This cousin proudly enumerated for my benefit the names of all the persons dependent on him for their livelihood. He justified their large number by referring to the many guests he had to entertain and whom he had to receive in the proper fashion, that is to say, surround them with a great deal of service. He belongs to a newly rich Bhathela family and has married a Desai woman. I, for one, willingly fell a victim to the catering from his kitchen. The food he served was more varied and of better quality than that of any other Anavil in the village. The fact that he is the only one to possess a *savari galli*, a vehicle used exclusively by prominent Desais for the transport of persons, also testifies to his eminent position. It is in his lavishly furnished house, not in that of the village headman, that important receptions and meetings are held.

The headman is the opposite of his cousin in many respects. A fairly young man whose political ambitions go beyond Chikhligam, he is a member of the Congress Party, for which he performs many tasks in the *taluka*. Besides being a landlord, he is a contractor, and through his connections in the capital of the subdistrict he has received various lucrative commissions. His preoccupations often take him outside the village, while his cousin stays at home. He has arrived at the conclusion that Chikhligam as a base is so small and so far from the center of power that it lacks scope for full development of his leadership. At the time of my departure the village headman was thinking of following the example of a rich Anavil landlord from a neighboring village who moved to the *taluka* town. From there he could regularly come by car to Chikhligam to supervise his property and interests. He was the first to try to do away with the perquisites to which his servants are traditionally entitled, which he proposed to replace by money payments. As yet, however, his plan has not succeeded.

Individual differences aside, all Anavils who have one or more Dublas in their employ still present, more or less, the traits of a patron of times past, if only because they are sometimes addressed by their subordinates in this role. The older servants in particular endeavor to play upon their master's liberality and high position by a show of servility that young Dublas think revolting. Furthermore, the landlord is expected to maintain his dignity under all circumstances. Brutish behavior is condemned by all, including his fellow caste members. On the other hand, the Anavil wins admiration when he has a natural authority over his subordinates, when he is tolerant and yet not to be trifled with. The servants can also count on the intercession of their master when they get into difficulties, usually over drink. The Anavils, however, do not concern themselves unduly with their Dublas. Insofar as members of the dominant caste still admit to having "obligations" towards their subordinates, they do so to be assured of an adequate supply of labor rather than from a desire to aid and protect the servants who depend on them.

Aside from a few individual exceptions and some vague notions of duty and honor that remain from an earlier period, a businesslike attitude of the landlords towards their agricultural laborers has come to prevail.

THE RELATIONSHIP BETWEEN ANAVILS AND DUBLAS. The Dublas in Chikhligam feel that the Anavil Brahmans have abandoned them to their fate, and are embittered by the realization that the landlords no longer need them. *Khavati*, formerly resulting from the responsibility of the *dhaniamo* for his subordinates' welfare, has increasingly come to represent an advance payment on labor not yet performed and is considered a favor that can be arbitrarily withheld by the master. Prior agreements among the members of the dominant caste concerning the amount of wage they are prepared to pay have resulted in more uniform treatment of the agicultural laborers. The landlords have become impervious to flattery, and appeals on the part of the Dublas to their magnanimity are usually in vain. The agricultural laborers, on the other hand, when they are paid the minimal remuneration, know that they are being deceived into the bargain. According to a report on their situation, a check of the servants' daily grain allowances showed 10 to 25 percent short weight.[20]

Have the developments of the last few decades increased the poverty of the Dublas in Chikhligam? A clear-cut answer to this is difficult, for it depends partly on how their condition in the past is evaluated. The Dubla farm servants themselves believe they are worse off; they extol the liberality and justness of the masters of the

20. *Report of the Hali Labour Enquiry Committee*, 27.

past. It is true that nowadays they are treated in a more impersonal way, but the stereotypes about the loving care they enjoyed in the past are not very convincing. Patronage does not exclude a precariously minimal livelihood for the clients.

Nevertheless I believe that the agricultural laborers among the Dublas have indeed become poorer. In the first place, the farm servants' social insecurity has increased since the disappearance of *hali-pratha.* They are now in an unstable category and their attachment no longer entitles them to any support, however slight; on the contrary, in old age and in case of illness they can be dismissed without much ado. Apart from this, the changeover to money wages has contributed to the deterioration of their condition. Insofar as the daily allowance is given in kind, it has remained at the same level for more than a century. Paid in money, this living wage has risen from 2 to 8 annas during the past fifty years and will no doubt further increase. But the prices of the articles they can buy for it have risen much more sharply. Instead of the emoluments in kind that the Dublas used to receive from their masters' gardens, they now are given a yearly bonus which, like their wages, is not adjusted to the sharply rising prices. Transformation of a subsistence economy into a money economy has taken place at the cost of the economically weak Dublas. Their wages are so low that even a box of matches must be made to last, although it costs 1 anna. Now that the tinder-box has fallen into disuse, they postpone lighting their *bidis* (country cigarettes) until the master comes to the field for inspection and can give them a light.

In contrast to the extreme poverty of the Dublas is the wealth of most members of the dominant caste. Nearly all of its members live in brick houses; only a few still have old-style dwellings of mud brick. Their rooms are furnished more expensively than in former times, with a sofa, cupboards, a clock, sometimes a table and chairs, and kitchen utensils of the coveted stainless steel. There is a wide array of "novelties," which the proud owners are eager to show. Men, women, and children are better dressed now, and the food and other items they use are of a better quality, for instance, wheat instead of millet, refined sugar instead of *gur* (raw cane sugar), and cigarettes instead of the local *bidis*; and milk tea flavored with spices has found general acceptance among the Anavils. A wide gap between the standards of living of landlords and agricultural laborers has, of course, always existed, but in the traditional economy the difference was one of quantity rather than quality. Owing to the process of economic enlargement of scale, the consumption patterns of the two castes have sharply diverged.

In a more or less closed village community it is natural that the landowners should share the crop with members of other castes. Partly, however, because of the disappearance of local isolation, agricultural products have attained a higher market value. The material results of the changes in agricultural production were not used by the landlords of Chikhligam in the traditional way, i.e., in sharing the profit with their Dubla clients — a change of preferences to the disadvantage of the agricultural laborers, for the money earned in mango growing was invested by the Anavils in education for their children. Apart from this, maximalization of income has become an end in itself for them. Besides undiminished "conspicuous leisure" it is now "conspicuous consumption" that characterizes the life style of the Anavils of the village. They enjoy their newly acquired wealth and buy luxury goods previously unknown in the village community, thereby once more emphasizing their distance from the other inhabitants. The conclusion is that the changeover to a money economy should be understood and evaluated in terms of both production and consumption.

Economically the gap between landlords and agricultural laborers has become wider. Although the latter may not be much worse off than before in an absolute sense, they measure their poverty against the increased wealth of the former. The Anavils admit that the wages of an agricultural laborer are low, but claim that they cannot pay more, an argument which fits in with their complaints about the low productivity of agriculture. A second motive they adduce is that the Dublas do not need much. The kind of life that a member of the dominant caste leads, or is expected to lead, requires a high income. The Anavils have become more cash-minded during the last few decades and in the calculation of cost and benefit of a farm servant, the profitability of his labor is now the chief criterion.

> It is obvious that the average of wages paid to the Hali is higher than that paid to the free labourer, while the actual output of work per day by the former is only equal to, if not less than, that of the latter, as the former, unlike the latter, is an irresponsible fellow. Thus evidently the Hali system is uneconomical and inefficient.[21]

It is interesting to note that the changed interpretation of the use of Dublas led at the same time to a change in stereotyped concepts

21. Mukhtyar, *op. cit.*, 170. See also Shukla, *op. cit.*, 130.

about them. Whereas writers of the nineteenth century had termed them hard-working, honest, polite, and hospitable, they are now looked upon as lazy, dishonest, and impertinent.[22]

At the time of my stay in Chikhligam a landlord dismissed a farm servant who was "overwise"; he did not know his place and always replied insolently when his master browbeat him. The Anavils complain that nowadays the promises of a farm servant are not worth a thing. Their proverbial devotion to the master's interests has been replaced by indifference and unreliability as the relationship between them has become more impersonal. A farm servant sometimes stays away for a whole day without notice, and should there be a dispute, he is likely to go off to the brickyards. "Like a hare, a Dubla does not remain long in the same place," is a proverb which the landlords like to cite. The lack of understanding among laborers is annoying, say the Anavils, their responsibility is nil, and their idleness knows no bounds. They are good-for-nothings who do not want to work so long as there is still something to eat. It seems likely, however, that it is not the performance of the agricultural laborers that has changed, but rather the weight given it by the landlords. While minimizing their obligations they now demand a maximal work performance, an effort which the *halis* were never asked to make in the past.

The Anavils are especially indignant that nowadays they sometimes have to go to the huts of the farm servants to ask them to come to work. Such a forced journey to the Dubla's quarter particularly annoys them. It is improper and far beneath their dignity, they feel, and they consider that they should be approached by the landless, not the other way round. For people who have been accustomed to being begged for employment, this is indeed a distasteful symbol of changed times.

As for the Dublas, they discovered, often the hard way, that the Anavils no longer wished to behave like patrons. No longer could the Dublas claim total care and generous aid as clients. "If the master's work does not suffer in the event of the Hali's sickness, he may be allowed to rot in his cottage," Shukla already noted in about 1930.[23] Although we may well wonder whether in the traditional community the master treated his *halis* as well as he claimed to do, a completely different relationship has surely come into being when the value

22. *Gazetteer of the Bombay Presidency*, 1882, vol. XIII, pt. 1, 158, reports favorably on the Dublas. For a negative judgment, see, for instance, Mukhtyar, 1930, 167–68, and *Report of the Hali Labour Enquiry Committee*, 1950, 45.

23. Shukla, *op. cit.*, 123.

of a farm servant is measured by his output, not by his behavior as a client. The landless laborers feel unprotected and exploited. They have seen the prosperity of the landlords increase and are aware that this has been at their cost.

Formerly the son of the farm servant grew up in the house of the master where his mother worked. He played with the son of the Anavil, who later became his boss. This intimate relationship between master and servant has now ceased. Subservience and respect are still demanded from the servant, but the personal tie that mitigated the inequality is no longer there. The landlords do not trust us, say the Dublas in permanent service. They give directions even when it is not necessary, only to make their authority felt. Never does a master say he is satisfied; whatever the farm servant does, the landlord always has some fault to find. Running away under cover of night, for the landlord a proof of faithlessness and dishonesty, is for the farm servant the only way to escape the authority of his master. To ask for permission is to be refused and to give the master an opening to exert pressure so as to prevent his servant from carrying out his plan.

That they depend on their landlord in everything is drummed into the minds of the Dublas. *Khavati* on non-workdays must be literally begged by the servant, and he does not get it until he has done some odd jobs around the house of the Anavil. Sometimes these jobs are clearly unnecessary, and are only meant to show who is master. The servant tries to avoid this confrontation by sending his children to his master's house to get the daily allowance or perhaps ask for *khavati*.

As the relations between the families of master and servant became more impersonal, the humble character of the tasks performed by the Dublas received more emphasis. It is the most disagreeable jobs that fall to them, and this applies even more to their wives who work in the Anavil houses. The women are often on bad terms with the mistress, who leaves the unpleasant work to them and has assumed more and more airs as the gap between them has widened. Among the Dublas there is a growing awareness that the low esteem they enjoy in the village is closely connected with the kind of work they do. The weak social and economic position of this tribal caste is, however, perpetuated by their illiterateness. Whereas the Anavil boys and girls go to school, very few Dubla parents have their children educated, and then only in the lower forms, to have them out of the way in the daytime. At the age of eight or nine the boys work as *govalia* and the girls look after their smaller brothers and sisters in their mother's absence. When the family goes to Bombay, the children contribute to its income by working in the brickyards.

Without wishing to keep up the former relationship, both parties appeal to their traditional rights while rejecting the obligations that went with them. The Anavil Brahmans no longer treat their laborers as clients, but still, and as a matter of fact, expect gratitude and meekness. On the other hand, the Dublas, especially the older people, still occasionally try to approach their landlord in his quality of patron. Because they have always been accustomed to receive a living wage, they do not recognize the recoverable character of advance payments. Nowadays, however, they try to avoid situations in which they are committed to unconditional obedience and loyalty. A relationship of attachment, formerly much valued because of the social security it offered, is now rejected. This is partly because the servant can no longer count upon the support and protection of his master, but above all because the Dublas regard the bondage inherent in attached labor as unacceptable.

For a Dubla with a family, it is impossible to live on a day-laborer's income, and this is the sole reason why some of them accept permanent service. But in order to reduce their obligations and to be free to sever the tie at a moment's notice if necessary, they try to pay for their marriage without the master's help.

Among the members of their caste, the farm servants are of little account. In the present situation, the recognition of dependence offers few advantages, while the disadvantages are many. Nowadays no Dubla girl likes to marry an agricultural laborer who is in permanent service, for she probably will have to work in his landlord's house and perform all the humble and dirty tasks that the Anavil women look down on. The farm servants are no longer a privileged category who are proud of their relationship with an Anavil. It is the migratory laborers whom the Anavil Brahmans have in mind when they say that the Dublas have failed their former "benefactors." Labor discipline in Chikhligam is said to have disintegrated through the irresponsible behavior of the seasonal migrants. Particularly the young people are looked upon as insolent, that is to say, they no longer display any humility in the presence of Anavils and no longer allow themselves to be abused, let alone thrashed, a treatment that in the past many a Dubla had to endure. Those remaining at home are led astray by the drinking and extravagant spending of the migratory laborers, say the Anavils. In the brickyards the Dublas develop a craving for various luxuries — they drink milk tea, for instance, and smoke real cigarettes, and one of them even wore a nylon shirt when he returned. In short, there they acquire habits to which members of the dominant caste feel that they alone are entitled.

Because of general dissatisfaction with the present situation, both parties tend to idealize the past. On the other hand, it cannot be denied that relations between landlords and agricultural laborers have deteriorated in recent years. The mutual distrust and antagonism of which I sketched the sources still prevail. The Anavil Brahmans fear the rise of the lower castes. The influence of the Dhodias increases, as I described earlier in this chapter. The rapid demographic growth of the tribal castes heightens the anxiety of the members of the dominant caste. They feel deluged by the *kaliparaj*, whom they believe inferior. The traditional power elite of the villages in south Gujarat resent that the government promotes the interest of these backward castes, at least in theory. In their opinion it would be much more useful if the authorities directed the attention of the agricultural laborers to their obligations towards their employers and taught them to respect their betters. That the members of the younger Dubla generation fail to show subservience is a source of boundless irritation for the landlords. They feel hardly able to hold their own in the new situation, and moreover their number in the village is diminishing. At present their caste is still dominant, and quarrels between caste members are smoothed over for the sake of unity. As they have come to resemble each other in their life style, Desais and Bhathelas are no longer strictly separated. Especially in the presence of outsiders, the earlier rigid distinction between the high and the low section of Anavil Brahmans has become blurred. They all speak of "we Anavils." Their feeling of solidarity has intensified under the threat presented by the low castes.

From the Dublas the Anavils have little to fear as yet. Even when Dublas are present the landlords speak rudely and offensively about the failings they attribute to this caste. By removing to a new village quarter, about half of the Dublas have managed to extricate themselves to some extent from the power of the dominant caste. When the land settlements were in preparation, the Anavils feared that the agricultural laborers would claim title to the land near the village pond on which their huts stood. The Vashi owners allotted other places on uncultivated and rough *jarayat* to the Dublas concerned, but the laborers refused. The affair was brought before the government authorities, since Dublas who wish to buy land for their huts can obtain a subsidy from the government. In this manner an important means of pressure is taken away from the landlords, who, in case of conflict, can summarily evict a laborer and make him homeless.

When in this case the Dublas in Chikhligam appealed to the government, it was ruled on the district level that they had to relinquish the land on which they lived. After this victory the Vashis also refused to accept the Dublas as tenants. A Bania from the *taluka* town offered to sell them a piece of land just outside the village which he would have lost anyway under the Tenancy Act. For the land and the twenty-eight houses to be built on it, the authorities gave a subsidy of 21,000 rupees, 750 rupees per hut. The Dublas had to contribute 250 rupees each in the form of labor, as the subsidy had been granted on condition that they build the houses themselves under the supervision of a contractor. But even I could see that the value of the material supplied to the Dublas was far below the allotted amount. The houses have numerous defects. It is their own fault, say the Anavils, that "the Dublas now have been swindled by officials of the subsidizing body as well as by the contractor. They should have asked our advice." For the Dublas, however, it made little difference whether they were swindled by officials or deceived by the Anavils, who would have done so if they had had the opportunity.

Just as expected, the Anavils' manipulations in this affair to prevent the agricultural laborers from obtaining the land to which they were entitled had the support of the regional bureaucracy. As evidenced by this example, the laborers now find themselves in an isolated position. They must still communicate with the authorities through the landlords, by whom their claims are put forward unconvincingly, if not damaged. Nevertheless, the landlords have not been able to prevent these Dublas from gaining a small amount of independence when they settled on their own land in the new quarter.

The migratory labor of the agricultural laborers has affected the Anavils' influence still further. The landlords' complaints about the faithlessness of the Dublas spring from rancor about their own declining authority rather than from any anxiety about an imminent shortage of labor. For every year at the peak of the agrarian season, and after a number of years permanently, most migratory laborers, however discontented and "insolent," are compelled to put themselves in the landlords' hands again.

Apart from the new possibilities of escape the Dublas occasionally resort to time-honored means of resisting the pressure of the Anavils. They can play upon the Anvils' fear of witchcraft, which can be traced to the tribal background of the Dublas. Some of them are renowned for their magic incantations, and the Anavils are among those who fear their sinister power. If the worse comes to worst, they can still escape from the landlord's control by running away to the village of a relative, usually that of a wife or a mother's brother.

The means of resistance available to the Dublas are not impressive. They are the second-largest caste in Chikhligam, but their power is not in keeping with their number, and politically they have not as yet been activated. The functionary of the C.P.I. (Communist Party of India) in the *taluka* capital devotes his efforts exclusively to the small farmers in the region. The Congress Party is so much a stronghold of the dominant caste that the caste of landless laborers does not receive much attention.

Although their migration to the brickyards near Bombay has heightened the Dublas' awareness of their position, they lack the means of changing it. In the brickyards they are removed from the urban sphere, uninteresting for trade unions and political parties, and uncanvassed. They leave the brickyard areas only for brief visits to a tea-stall, a cinema, or another brickyard where fellow-villagers and relatives are employed. Thus far no Dubla leaders have come to the fore to give guidance to the protest of their fellow caste members against the low social position and economic distress in which they live. Moreover, the level of education in this caste is so low that there is no prospect of an early change.

All in all, the authority of the caste of landlords over that of the agricultural laborers in Chikhligam is still considerable. The only way of protest open to the Dublas is that of patient and unremitting resistance. Hate and enmity towards their exploiters remain below the surface, and in the presence of a landlord these feelings are rarely evidenced. On the whole, the laborers behave in a lethargic and indifferent fashion, avoid giving answers, know nothing and can do nothing, take no initiative, and accept no responsibility. The inertia of the Dublas has become proverbial. For the Anavil Brahmans this mechanism of passive resistance is merely another proof of the Dublas' inferiority.

Chapter 8

Gandevigam

On whose head did I see a green basket?
I saw it on the head of Dolto, the son-in-law of the family.
In whose stable did I see him picking up cowdung?
I saw him in my brother's stable.
I took him to be a Khalpo and beat him with shoes
I took him to be a Dhed and pelted him with stones
I took him to be a Dubla and beat him with my sticks
He said, "For God's sake do not beat me,
I am Khandubhai's son-in-law."
T. B. Naik, *Songs of the Anavils of Gujarat*

Leaving your father's place
Why have you come here?
Your father-in-law has a ramshackle hut
Your father-in-law is a serf of the whole village
Why have you come here?
Your father is an honourable man in the village
What brings you here?
T. B. Naik, *Dubla Songs*

8.1 Region, Village, and Population

Gandevigam is one of the smallest subdistricts of south Gujarat. This Gandevi *mahal* forms part of the central plain, an ancient culture area with intensive agriculture. The subdistrict has always exceeded the others in soil fertility and population, as early government reports show.[1] The urban center of Gandevi has long served the surrounding countryside. Agricultural products from the villages are marketed here. The town's situation on a navigable river not far from the coast formerly facilitated shipment of goods

1. See the first survey and settlement report of Jalalpor *taluka* of 1868, included as appendix R in the revised report dated 1900: "Gundevee, a purguna belonging to the Guicowar, and famous for being the most fertile as well as the highest assessed in the Western Presidency," 1868, 1.

from and to important ports, especially Surat, the large seaport on the west coast of India. The smaller population centers for miles around availed themselves of the facilities of trade and the artisanal activities that flourished in this harbor principality in its heyday. Its sphere of influence, no doubt, also comprised the region of Gandevigam, and it is not impossible that the ancestors of the present village population contributed to the food supply of the city, which, by the end of the eighteenth century, had about 600,000 inhabitants.

The political turmoil that led to the decline of Surat ushered in a time of stagnation for the whole of south Gujarat. The vicissitudes of war between the Moghul forces and the Marathas brought insecurity and economic distress. Bands of robbers ransacked the countryside, and the population long remembered, in particular, the plundering raids of Shivaji, the Maratha chief. Even now, certain villagers can describe how and where their ancestors used to hide the crop and other possessions at his approach. Profiting by the political vacuum, men wielding local power levied such high taxes that a large part of the land remained uncultivated and whole villages were depopulated. This may explain why in a European travel record of the late eighteenth century the region round Gandevigam was described as unreclaimed.[2]

After many decades the struggle for power was decided in favor of the English and the prince of Baroda, a vassal who had managed to free himself of the Maratha rule. They divided the territory of south Gujarat between them, and Gandevi *mahal* fell to the principality. The town remained a center of local administration. It was the seat of a very prominent Desai family which had long ago received from a Moghul prince various seignorial rights over a number of villages in the vicinity and which now pledged loyalty to the new ruler. The influence of this family of landed nobles declined sharply in their traditional domain when, in the second half of the nineteenth century, the system of revenue farming was abolished, although they continued to hold a prominent position as large landowners.

The political and administrative stabilization that took place during the early nineteenth century did not lead at first to an economic revival. The river mouth silted up and Gandevi town became

2. "Although the soil is very favourable about Gaundevee, I saw but little culture: the greatest part of this district, as far as the eye could reach, is one field of high pasturage." (Hove, 1787–88, 93–94.)

inaccessible to shipping, and consequently artisanal activity deteriorated even further. Only after the north-south railroad was constructed around 1860 did commerce and the economy in general begin to expand again, though mainly in the new towns along the railroad. In population and economic importance Bilimora came to far outdistance Gandevi town. Whereas in 1881 it was an insignificant little railway village with fewer than 5,000 inhabitants, Bilimora has since grown into a town of over 30,000, with some textile mills, a paper factory, and a number of small industries in the metal sector. It is also a center of trade in timber, which is brought out from the woods in the east over a narrow-gauge railway.

Although Gandevi town was connected fairly soon with the railroad by the building of a highway to Bilimora, it remained at a disadvantage on account of its offside location in the new communication network. Its population long experienced no growth, and early in the twentieth century a plague epidemic reduced it by about one-fourth. Since then the population increased very gradually, and after Independence with greater rapidity, to 11,000 in 1961. The building of a cooperative sugar factory, the promotion of textile manufacture in small enterprises or as domestic industry, and the extension of local government services contributed to this growth. Gandevi has remained the administrative center of the subdistrict and has an economic function as distribution point for the surrounding villages, including Gandevigam.

After Independence, the principality of Baroda was incorporated into the State of Bombay, later Gujarat. The various parts of the former principality were merged into the local administrative structure. Gandevi *mahal* was included in the Surat district, which was divided into two parts in 1963. Like Chikhli *taluka,* Gandevi *mahal* is now part of the Bulsar district.

The subdistrict is known as a very prosperous agricultural area. It owes this reputation chiefly to the fertile soil found along the upper reaches of the Ambika River to the northeast of the town. This region is called the garden of south Gujarat, which is the more striking as the whole southern part is one of the most fertile areas of Gujarat. The great fertility of the soil along the river has led to the high population density of 916 per square mile in 1961 for the subdistrict as a whole, a figure far above the average of Gujarat and exceeded by none of the subdistricts in the southern part of the state. Within a radius of less than two miles from the village nucleus, Gandevigam is surrounded by five other villages, which is an indication of the extremely high concentration of population in the area. This is the first important contrast with Chikhligam. There is little to distinguish Gandevigam from the other garden villages in their rural surroundings and, as we shall see, they are closely interconnected economically and socially. Soil conditions are noticeably different from those in Chikhligam, where much of the surrounding land is, at present, considered uncultivable. In contrast to the arid and barren aspect of that landscape in the summer and winter months, the much more intensive agriculture in the garden region gives the land around Gandevigam a lively and pleasing appearance. Mango orchards, cane fields, rice fields, beds of vegetables, and banana plantations make an attractively diversified picture; compared with this variegation of color, the rural scenery in the vicinity of Chikhligam is monotonous after the monsoon ends. Finally, being situated in a more urban and economically more developed subdistrict than Chikhligam, Gandevigam is easy to reach. The village borders on the asphalt road that runs from Bilimora by way of Gandevi to Navsari, the chief town of an adjoining subdistrict. In the daytime there are buses in both directions every two hours from the village stop on the road. The distance to Navsari is eight miles, to Bilimora seven miles. Halfway between the village and Bilimora the road passes the town of Gandevi at three miles' distance.

I have not been able to learn the age of Gandevigam with any certainty. Its site in an ancient culture area suggests a much longer history than that of Chikhligam, but the village inhabitants could not verify this. They say that the village was founded long ago by a shepherd who had a garden in this spot. The first Anavil Brahmans to settle here were two brothers, to whom some families still trace their ancestry. Gandevigam was an *inamdar* village. The prince of Baroda had ceded land titles and a number of other prerogatives, not police rights, to an *inamdar* for services rendered. This *inamdar* lived in Baroda, where he was in the maharajah's employ, and visited Gandevigam once or twice a year. He was represented in the village by a steward. After the annexation of the principality in 1950, the rights of the *inamdars* were annulled. Since then, the land revenues no longer go to these large absentee landholders, whose main privilege it was, but the farmers pay taxes to the government. For the village population this change had few, if any, consequences. The steward still lives in the village and has received land from the *inamdar* to assure him a living.

In 1963 Gandevigam had slightly more than 1,000 inhabitants, according to information given me by the local authorities. With a few exceptions the villagers are members of three castes: Anavil Brahmans, Kolis, and Dublas. When I arrived at the village the first thing I saw was the huts of the Dublas on both sides of the asphalt road. A large part of the village territory lies between this road and the Ambika River. Where the two meet is the village nucleus with the houses of the other inhabitants. There are two Anavil streets and three Koli streets, spatially separated but connected by paths. The rows of houses are behind trees and invisible from the road. A little farther along, wedged in between the road and the river, are the gardens, nowadays mainly planted with fruit trees. The land across the road is given over to completely different crops: rice, bananas, vegetables, or sugar cane according to the season. The dense vegetation makes for a relatively temperate microclimate in the village territory even during the hot season, a circumstance which I, as a West European, highly appreciated, especially in the afternoon hours. In the shaded *bagayat* villages the temperature is always a few degrees lower than in the more open *jarayat* villages of the plain.

In the Anavil streets many houses are empty for the greater part of the year. It is characteristic of the villages in the *bagayat* region that during the dry winter and summer months the Anavil Brahmans live in their gardens, where they are surrounded by the crops they cultivate. They used to have small and simple wooden dwellings there, not much more than sheds. For in the narrow streets of the village nucleus, where the cattle usually remain much nearer the houses, insects are a veritable plague. To stay in the shaded gardens is far more agreeable, certainly in summer. During the rainy season, however, the narrow paths become all but impassable, and the Anavils were wont to return to their street before the monsoon broke. Increasingly, the wooden cottages in the gardens have been replaced by large two-storey brick houses, with spacious front and back yards and separate outhouses, and as these new abodes are much more comfortable, many Anavils remain there the year around, neglecting their permanently closed houses in the village street. I was lodged in one of these. When the family estate is divided, one brother sometimes gets the house in the village nucleus, the other the house in the garden. Several Anavils are too poor to have a house built in their gardens, especially as it means the loss of a parcel of the most fertile arable land. Nevertheless there is, generally speaking, a tendency within the dominant caste to take up permanent residence in the gardens.

The majority of the Kolis' houses are not built in rows but are scattered along the three streets they inhabit. Their houses are smaller than those of the Anavils, lack electricity, are built of wood and wattle instead of brick, and have roofs of tiles or corrugated iron. Each street has one well, whereas many houses in the Anavil quarter have their own well. The cattle are generally sheltered in the house.

The Dublas live along the asphalt road on the fringe of the village. Their small, low mud huts with overhanging roofs of straw or sugar-cane leaves are on rough and low-lying terrain. A few members of this caste are better housed: their dwellings are partly built of wood or brick and have tiled roofs. None of the houses, however, has more than one room. Although this Dubla quarter is of fairly recent date, and is the largest in the village, its inhabitants have only one well among them. Until the early nineteen-fifties their huts were scattered in the gardens of the Anavils who employed them. Concentration of the Dublas and a tendency towards deconcentration among the Anavils are therefore important changes in the residence pattern of the past few decades.

Facilities in Gandevigam are few. The stock of the two village shops, both run by Kolis, is confined to victuals and other daily consumer goods. One shop is situated on the asphalt road near the Dubla quarter and is exclusively patronized by members of this caste; the other is nearer to the village nucleus. Here, too, is the village school. For all other amenities the inhabitants depend on the neighboring larger villages or on the town.

As in Chikhligam, collecting data on Anavil Brahmans and Dublas was the main purpose of my stay in Gandevigam. In 1963 I counted in the village 185 Anavils distributed among twenty-six houses, and 381 Dublas in sixty-one huts. This figure is lower than that mentioned in the village records,[3] but the local government data are not wholly reliable. The steward who remained in the village, for instance, has not been registered as a Maratha Brahman but as an Anavil. A Dhodia and Naika family living among the

3. In these records, 225 Anavils and 429 Dublas are registered as inhabitants.

Dublas have been included in the count of their fellow inhabitants of the quarter.[4] I further learned that a number of persons who no longer live in the village are still registered there. Finally, it is uncertain whether some Dublas live in the territory of Gandevigam or in that of the neighboring village. They were probably incorrectly registered as inhabitants of Gandevigam in the latest censuses; when sugar was rationed during the Second World War, they had to get their ration cards in the other villages. How much lower my total count is I do not know, for I did not include the Kolis in my census. I have taken data from and about them only insofar as such facts could throw light on the relationship between landlords and agricultural laborers. As will become clear, they belong to neither category. Background information was given me by randomly selected Kolis, a sample of thirty-seven taken in three streets. On the other hand, I included in my investigation all Anavil Brahmans and Dublas who were present in the village. Where not expressly stated otherwise, the quantified data that will be discussed below have been taken from my own census.

8.2 Economic Position and Activities of the Inhabitants

The inhabitants of Gandevigam differ widely in prosperity, by and large according to caste. The disparate quality of their accommodation was mentioned above. Two years before my stay the richest Anavil of the village had a house worth 30,000 rupees built in his garden, whereas the value of most Dubla huts does not exceed 100 rupees. The middle position of the Kolis, better than that of the Dublas but worse than that of the Anavils, is also apparent in other respects. For instance, the average amount of money spent on a marriage, according to my informants, is for Anavils more than 5,000 rupees, for Kolis 1,000 rupees, and for Dublas about 300 rupees. It is true that the most prosperous of the Kolis are better off than the poorest of the Anavils, and the way of life of the poorest Kolis is very similar to the penurious existence of the Dublas. Nevertheless, a connection exists between living standard and caste membership, which appears most clearly from differences in landholding among the various castes. The Anavil Brahmans, as in Chikhligam the smallest group, control a disproportionately large part of the land. The Dublas, numerically the largest caste, are with one exception completely landless. In the preceding generation

4. Like the Dublas, the Naikas are a tribal caste in the plain of south Gujarat.

one Dubla received a plot from the *inamdar* for whom he had worked as a farm servant. On this plot, far into the gardens, are to be found the huts of his two sons and three other relatives.

TABLE 8.2.1

CASTE MEMBERSHIP AND LANDHOLDING IN GANDEVIGAM*

	Property		*Population*	
Caste	*Acres*	*Percent of total*	*Number*	*Percent of total*
Anavil	260	71	225	22
Koli	62	17	350	35
Dubla	0.25	—	429	43
Others	42.50	12	—	—

* Data obtained from the village records.

In the whole of south Gujarat the Kolis are petty cultivators, and those in Gandevigam are no exception. The members of other castes listed in table 8.2.1 are inhabitants of surrounding villages who possess land in Gandevigam.[5] Conversely, landowning villagers, especially Anavils, possess plots elsewhere, which illustrates the intermingling of the village economies in this *bagayat* region. The picture emerging from the record of land rights kept by the village clerk (*talati*) could be supplemented by the following survey, for which I myself gathered the data. These figures relate to holdings both within and outside Gandevigam. The inequality of the landowning castes in terms of separate households is even more pronounced when it is taken into account that the Anavils possess the great majority of gardens, i.e., the most valuable plots. Kolis own mainly less valuable *kyari* and *jarayat* land.

It is important to make a distinction between ownership and cultivation. Of the Kolis I interviewed, sixteen are sharecroppers of Anavil landlords. The land concerned is mostly *kyari*, on which they cultivate sugar cane, bananas, or vegetables. When they grow vegetables, the Kolis vacate the land by the end of the summer, so that the owner can grow rice on it during the monsoon. Landlords and sharecroppers each pay half of the cost of seedlings, fertilizer, and irrigation water, and share the produce equally. The crop-sharing system has enabled the Kolis to hold their own as

5. Some of these are also Anavil Brahmans, who together possess about thirty acres. The rest of the arable land in this category belongs to a Bania and some Moslems from a neighboring village.

smallholders. Tenancy, as well as sharecropping, however, is now officially prohibited, and the Anavils run the risk of losing their land. Now that crop-sharing arrangements no longer have a legal basis, it is not surprising that the number of sharecroppers among the Kolis has recently declined. Moreover, to limit their risks as much as possible, the Anavils now lease out the plots for one season at a time, which means great insecurity for the tenants.

TABLE 8.2.2

CONCENTRATION OF LANDHOLDINGS, BY CASTE, IN GANDEVIGAM

Landholding (acres)	*Caste of owners*	
	Anavils	*Kolis*
—	1	5
Less than 3	3	20
3 to 6	6	4
7 to 16	7	2
More than 16	9	1
Total	26	32[1]

[1] The information given by five Kolis was too vague to be classified.

A large proportion of the Kolis (thirteen of my thirty-seven informants) have to carry out field labor for others occasionally to make ends meet. The extent to which they succeed in doing this depends on the quantity of land they cultivate themselves and on their income from other sources. Two Kolis in Gandevigam depend entirely on their earnings from agricultural labor. The father of one of them was a tenant of an Anavil in a neighboring village until he died, and the informant himself was until recently a sharecropper of a landlord in Gandevigam. He has lost his team of oxen since then, and now he possesses nothing. He can no longer afford the education of his son, who went to secondary school. In the socioeconomic scale he has clearly lost ground.

Yet there is no reason to speak of a general deterioration within this caste. Ten of the Kolis in my sample have managed to add to their property under the Tenancy Act, though almost none by more than 1 or 1.5 acre. In all these cases the landlord lived outside the village. Then, too, more Anavil landlords now leave the plowing of their land and the transporting of the crop to these small cultivators. An increasing number of Koli families, finally, supplement the family income by the earnings of a member working permanently or temporarily outside the village. Thus, while the middle position of this caste as a whole has not fundamentally changed, a process of differentiation in economic activities and prosperity level has begun.

By far the majority of Dublas are laborers on the land, but some have lately been able to rent some land for sharecropping from Anavil landlords who did not want to lease it out to Kolis. It is to be expected that this trend will continue. The conditions under which the Dublas obtain this land, on which they cultivate bananas on a crop-sharing basis, are less favorable than those allowed to Kolis. The latter are entitled to half the produce, whereas Dublas receive only one-third. Although the landowner in the Dublas' case is obliged to pay all expenses, he often demands repayment of half of them when the accounts are settled at the end of the season. For such an intensive crop as bananas, some Koli informants said, crop-sharing under these conditions is not economic. A Dubla, who measures his earnings by those of an agricultural laborer, thinks otherwise.

Is there any reason to assume that in earlier times the land was distributed differently among the inhabitants of the village? The Dublas, I believe, have always been in the same circumstances. There are no indications that members of this tribal caste have ever been engaged in agriculture as independent cultivators.

For the Kolis, too, large-scale loss of land seems improbable. They were mentioned as sharecroppers and agricultural laborers in the employ of Anavil Brahmans as long ago as the mid-nineteenth century. Although several Kolis were dispossessed and became tenants on their own *kyari* and *jarayat* land in the nineteen-thirties, the Anavils also had to give up land in that period of world crisis, mostly to Banias (traders and moneylenders). In the ensuing years, many Anavils and Kolis could buy land again, and the situation of both castes improved considerably. During the Second World War, a time of scarcity of food crops, the cultivation of sugar cane and bananas in Gandevigam was extremely profitable.

Paradoxically enough, the Tenancy Act, which was meant to help the Kolis, will probably weaken the economic position of many of them. In practice this act has confirmed the traditional ascendancy of the Anavils as landowners in the village community.

As a rule, the Kolis need little extra labor on their land, for the assistance of members of the family, including women and children, is usually sufficient. At the peak of the season, collaboration with other Koli families is one solution. Yet, despite the pattern of small landholding within this caste, about half the households about which I collected data employ from time to time the labor of others against payment. Although the normal daily activities can be carried out by the family, outsiders must be taken on at harvest time, especially when the family does not wish to exchange labor with other Koli families — often a source of annoyance. One of the reasons for this periodic need of labor from outside the family is that Kolis grow highly labor-intensive crops, such as sugar cane, vegetables, and bananas, on land which a number of them still cultivate under a crop-sharing agreement. I found that Kolis employ mainly their fellow caste members for casual labor. This occurs not so much because they are inspired by a magnanimous wish to support members of their own caste, as some of them claimed, but because the services of the agricultural laborers *par excellence*, the Dublas, have largely been monopolized by the Anavils.

Most of the Kolis who hire labor incidentally are themselves employed as day-laborers from time to time. Understandably, as small farmers they give priority to working on their own land, but this means that, when they are free, other landowners have little to do either.

Koli sharecroppers were formerly obliged to perform field labor for their landlords if necessary. As already described, however, the Anavils enter into fewer crop-sharing arrangements with Kolis, and the Kolis, for their part, are less disposed to work for Anavil Brahmans in the fields, for they no longer feel obliged to help their former landlords.

As the tenancy relationship diminished, a small proportion of the Kolis have come to depend increasingly on income from wage labor for others. They prefer to work for fellow caste members who, they say, treat them better and pay a little more. The category for whom agricultural labor is the main source of income is not large. Most Kolis have remained small farmers who try to supplement their earnings by means of various agrarian and nonagrarian side activities. Some distill and sell spirits illegally, breed and trade in cattle, or are small-scale mango buyers who put up the fruits for auction in towns along the railroad against prices that are slightly higher than the local ones. Koli women leave Gandevigam early in the morning with baskets of vegetables on their heads to hawk their produce around in the villages or to offer it for sale in the markets of neighboring towns. Many Kolis are cart-drivers for members of the dominant caste and also, with their oxen, plow the landlords' ground for 10 rupees a day. They regard agricultural work for others as beneath their dignity and try to avoid it as much as possible. This attitude is prompted both by the low remuneration of this work and by the slight prestige that it carries. Although the average landholdings of the Kolis are small, few — not more than five of the thirty-seven households in my sample — are completely landless. Of these, three are sharecroppers. Possibly, as population growth continues, a higher percentage of the members of this middle caste will retrogress to an existence of landless laborers, but this is not very probable. An increasing number of Kolis manage to find work outside agriculture in the neighboring towns.

On the basis of landholding, therefore, and leaving individual differences within each group out of consideration, three categories can be distinguished in Gandevigam:

1. Anavils, who, as in Chikhligam, own most of the land and the most fertile part of it, and who moreover possess relatively more equipment, such as motor pumps, agricultural implements, and cattle. Most of them employ one or more farm servants permanently, and rarely work on the land themselves. It is mainly for this reason that I have classified them as landlords, not farmers.

2. Kolis, who as small farmers mainly possess *kyari* and *jarayat* land, and who supplement their income by a series of other activities. It is true that members of this caste sometimes call upon the labor of others, and themselves occasionally work for others. As a rule, however, these roles of employer or employee are temporary, and often they are combined in one individual.

3. Dublas, who largely earn their livelihood as agricultural laborers. This made it possible for me to concentrate my investigation into the relations between landlords and agricultural laborers on the Anavil Brahmans and Dublas in Gandevigam, as I had done in Chikhligam.

Thus far I have confined my remarks to activities and positions connected with the agrarian economy of Gandevigam. As already mentioned, the total income of Chikhligam is earned in large part outside the village and agriculture. It might be expected that this proportion would be even higher in Gandevigam, for the village is situated in the immediate vicinity of some urban centers, with which it is connected by an excellent road. But, as the census returns of 1961 showed, 88 percent of the working population is engaged in agriculture. As compared with Chikhligam, the proportion of inhabitants of the garden village who have nonagrarian sources of income is relatively much smaller. In view of the more isolated location of my first field village, at a greater distance from the cities and less easy to reach, the pronouncedly agrarian basis of existence among the population of Gandevigam is noteworthy.[6]

TABLE 8.2.3
NUMBER OF INHABITANTS OF GANDEVIGAM EMPLOYED OUTSIDE AGRICULTURE

Caste	*Living and working in Gandevigam*	*Living in Gandevigam and working elsewhere*	*Temporarily living and working elsewhere*	*Permanently living and working outside Gandevigam*
Anavils	—	4	—	9
Kolis[1]	4	8	6	13
Dublas	—	25	5	24

[1] The data on the Kolis relate only to those included in my sample. I estimate the total number of Kolis employed in nonagrarian occupations to be higher than that of the two other castes, both absolutely and relatively.

Within the village itself only four inhabitants are engaged outside agriculture: two Kolis are shopkeepers, one is a carpenter, and the fourth is a tailor. Aside from these occupations, they own a plot of land and partly depend on its produce for their maintenance. The reverse is true of some of their fellow caste members, who are primarily small cultivators but add to their earnings by such activities as dealing in wood and illegally distilling spirits.

The proportion of inhabitants who live in Gandevigam but work elsewhere is larger. A small category, consisting entirely of Kolis and Dublas, leaves the village for regular periods and returns before the rainy season.

Many inhabitants may have gone to live in the neighboring towns, and their definitive departure would render the proportion

6. Cf. the continuing and even increasing stress on agriculture which T. S. Epstein found in Wangala, and her contrasting findings in Dalena, the other village comprised in her field work. (Epstein, 1962, especially ch. II.)

of the village population working outside agriculture much higher. I therefore inquired about those who had left the village in the past few years.[7] Once more, Chikhligam had the larger number of former inhabitants who had permanently departed. Thus, while in the vicinity of Gandevigam much more alternative employment is available than in Chikhligam, relatively few inhabitants make use of it.

My data show that, outside agriculture, Anavil Brahmans are engaged in the highly skilled and best-paid professions, and most often they are government employees. Dublas are largely unskilled and perform low-paid work. A job with the government or in a factory is much coveted by them, as it means a fixed income. When they succeed in obtaining such a job they get only the less-favored tasks, such as cleaning and maintenance work. While the Anavils from the village who work for the railway are employed as station-masters or ticket-clerks, Dublas from Gandevigam are luggage porters, watchmen, track walkers, or car cleaners. There are few artisans among them: two tinkers (father and son) and one brick-layer.

The Kolis who live in Gandevigam but work outside the village are, except for one teacher, carpenters or bricklayers. Six members of this caste leave the village periodically to work in an umbrella factory in a town north of Bombay. They leave their wives and children in the village, unlike the Dublas in this category, who migrate to the brickyards on the fringe of Bombay. Their number is much smaller than in Chikhligam. The figures show that Kolis, too, perform more skilled and better-paid labor than Dublas. The Dublas who work outside the village total twenty-five, of whom two are itinerant tinkers who travel about the surrounding country-side and three are timber cutters for a merchant in a nearby village, where two other Dublas work in a grain mill; and after the rainy season, one Dubla goes daily to a brickyard at a short distance from Gandevigam. Of the other seventeen, who work in the capital or in Bilimora, three are hands in a paper factory, three roadmen, six servants in shops or workshops, seven bricklayers, and four transport workers. Many of them walk the 3 miles to their work in Gandevi every day, and some even go 7 miles on foot to Bili-

7. As in Chikhligam, the statements are only approximations of reality. The longer the time since the departure from the village, the greater the chance that the inhabitants will forget those who departed, especially when no close relatives have remained behind.

mora. Anavils and Kolis who work in these towns all travel by bus or by bicycle. In what way the Dublas who permanently live and work outside Gandevigam are employed I could not find out, except in a few cases. Some work in factories or as railroad workers and sweepers, and some are vegetable growers in Bombay.

Obviously, Dublas who have had little or no schooling are not considered for skilled and well-paid jobs. Yet education is not the only factor in the selection of employees for nonagrarian occupations. As in Chikhligam, alternative employment has come to mean something different for each caste. The position in the village community and the economic and social content of caste membership determine to a large extent the kind of work that is considered suitable outside it. This confirms my findings with regard to Chikhligam.

8.3 Caste Hierarchy and Political Structure

The small extent of caste differentiation is the most striking feature of the social structure of Gandevigam. Economically, politically, and ritually the Anavils, who all belong to the Bhathela section, are at the top. They are followed by the Kolis, and the Dublas are classified as lowest. That is to say, high, middle, and low are represented in this village by only one caste each.[8] Curiously, Chikhligam, situated in a frontier area, contains more castes than Gandevigam. For a village in the central plain one would expect a more varied constellation. Serving and artisan castes, whose presence would suggest a division of labor characteristic of the Indian village, are completely absent. Perhaps it is precisely its location in a more developed region, with urban centers nearby, that contributes to the simplicity of stratification in Gandevigam. The villagers all belong to castes that, originally, were decidedly agrarian. Nonagrarian castes may have been represented in the village at an earlier period. At the beginning of the twentieth century there was mention of migration of artisans in south Gujarat to the towns as a result of the disintegration of the closed economy,[9] but I found no

8. Aside from a few households, which have erroneously not been registered separately in the village administration and the census report.

9. "The complete equipment of artisans and menials with which the old type of village was furnished is being dissolved by the force of competitive tendencies. As villages become larger, the village barber, blacksmith, carpenter or potter seems to lose the indefiniteness of his circle of clientele. The influence of custom in fixing the remuneration for the hire of labour is also giving way gradually to the laws of supply and demand. In many important directions the village services are being

indications that this applied to Gandevigam. More probably, a number of requirements in this area have always been supplied by urban centers. For instance, the rough clothing of the tribal castes was traditionally woven in the nearby town, whereas the demand of the inhabitants of Gandevigam for services of all kinds is mainly filled from neighboring larger villages, which contain various artisan and serving castes.

Even so, however, there is no question of a profound division of labor. As in the whole of south Gujarat, the Dublas in Gandevigam carry out widely varied services for the Anavil Brahmans, whereas the Kolis are largely autonomous and do their own cleaning, scavenging, and the like. When more specialized tasks must be done, they call in outsiders; for such services as those of a priest, hairdresser, blacksmith, or potter, they seek members of those castes from neighboring villages. The higher their own position in the caste hierarchy and the more prosperous the household, the more likely they are to call upon specialists. This means that it is primarily the most prosperous Anavil Brahmans who act as typical *jajmans*.

The calling in of specialists is made possible by the short distances between the various settlements in this densely populated region. It follows that Gandevigam cannot be studied as a separate social system. Only at the supralocal level has caste hierarchy a real meaning.

That Gandevigam forms part of a regional community also finds political expression. To obtain an insight into the local power relationships it is, therefore, not primarily the village system that is relevant. Athough the Anavil Brahmans are less numerous than the two other castes, they have three representatives in the village council as against two each of the Kolis and Dublas. The overrepresentation of the highest caste in the *panchayat* is a reflection of their ascendancy. Moreover, both the village headman and the police *patel* are Anavils.

The supremacy of this caste in Gandevigam is undisputed but means little. In the first place, decisions are not made in the village council. The *sirpanch* — village headman and chairman of the council — has no influence at all in Gandevigam. He is an aged bachelor who lives on his pension as a teacher and on the income

depleted by the discontented village artisan or menial leaving for towns or large centres in the hope that with better wages and in newer surroundings his ambitions can be satisfied." (*Census of India*, 1921, vol. XVII, pt. 1, 358.)

from his small mango garden. Nor are the other Anavil councillors influential personalities. This is not due so much to factionalism with behind-the-scenes activity as it is to lack of involvement among the Anavil Brahmans in the village as a social unit. They regard themselves not as inhabitants of the village but as *bagayat* landlords who live in the region extending on both sides of the river. This whole area on the Ambika is the domain of the Anavil Brahmans, or more precisely, of the Bhathelas. The dominance of this group is not confined to their functions in the village system, but affects also the regional frameworks of integration, such as the secondary school and the agricultural cooperative.

Together with the other children of the neighborhood, the boys and girls of Gandevigam attend the secondary school that has been built amidst a number of villages. In the past few years an increasing number of pupils come from the middle castes, notably the Kolis. Yet it is Anavils who rule the roost in the school board, and a disproportionately large number of pupils still belongs to this caste. This is even more true of the teaching staff. Some years ago, a Koli from Gandevigam succeeded in becoming a teacher at this school. However, as his brother told me, owing to the opposition and jealousy of the Anavils his situation became untenable. The Anavils did not think it proper that a Koli should hold such a post, above all at their school. In the end, he emigrated to East Africa.

In the agricultural cooperative, the dominance of the Anavil Brahmans is even more evident. The storage shed and office of this institution are situated on the asphalt road just outside the territory of Gandevigam. This cooperative has a pronouncedly regional function, not only economically but also socially. The Anavils of the village do not meet each other in the village nucleus. In the afternoon they emerge from their houses and set out for the site of the cooperative, where they congregate and meet their fellow caste members from other villages. Koli members go there only for business, not to have a chat. Upon my arrival in Gandevigam, I was not taken to the village headman but directly to this meeting-place. When, after my introduction into the circle, the question of my lodgings came up, this at first appeared to present some difficulties. But any objections were put aside as irrelevant by all those present. Even if I were to persist in my plan of exclusively collecting information about farming as it was done in Gandevigam, they thought it did not matter where I stayed. Not only because distances were short: I could daily meet the landowners of Gandevigam (meaning the members of the dominant caste) at the cooperative building.

The more than 700 members forming the cooperative live in nine villages in the neighborhood and belong to twelve different castes. Nevertheless this institution is entirely an Anavil affair. All of the fifteen members of the board are Anavils. The Koli members from Gandevigam feel discriminated against. The Anavils are, as it were, preferential members who can obtain credit more rapidly, on easier terms, and in greater amounts. This is no doubt partly due to the smallness of the Kolis' holdings, and in general to their dubious solvency as small cultivators. As a rule, large farmers everywhere have more influence in most cooperatives. It is quite likely, however, that Kolis are systematically put at a disadvantage for noneconomic reasons, and especially on account of their low position in the local hierarchy. The less prosperous Anavils benefit by belonging to the dominant group. They enjoy numerous advantages because of their caste affiliation, whereas the richer Kolis have an inherent handicap. Their stronger economic position in comparison with that of the least well-to-do Anavils is offset by their lower social status.

Because of its density of population, the whole region forms an ecological unit whose inhabitants know that they are closely interconnected socially and economically. This impression is strengthened by the regional control exerted by the Anavil Brahmans. Twice within a short time, during my stay, the chairman of the cooperative gave an expensive wedding feast for a daughter, and on both occasions sumptuously entertained many hundreds of guests from the environs. By this exploit, informants assured me, he stood a fair chance of becoming one of the ten most prominent Anavils of the region.

The background of the leading role played by the Anavil Brahmans must be sought in their privileged position at the time of the principality of Baroda, when members of this caste had administrative rights as revenue farmers. The system of revenue farming was abolished in the principality by the end of the nineteenth century, later than in the British area. Although the abolishment caused their powers to be curtailed, the caste as a whole retained its proprietary rights to the largest part of the arable land. This fixed the basis for a continuance of power, which was exerted by the members of the Bhathela section as well, even when they had not fulfilled any administrative functions on a regional level. As everywhere, the Anavils of the *bagayat* region are accustomed to being obeyed. They act with assurance and are described by the other castes as hard, cunning, avaricious, and rude. Such pejorative stereotypes also figure in the proverbs that circulate about them and that contain warnings about their greed, untrustworthiness, and rude language.

During my stay there was a great commotion in the adjoining village, where a member of the caste of Blacksmiths was anonymously threatened with a nocturnal attack on his house. According to local gossip he had earned much money with his engineering workshop and by dealing in mangoes. This had apparently provoked the jealousy of the Anavils, who were even more indignant when on the occasion of his daughter's wedding he held an expensive feast. It was a public secret that the threatening letter came from some members of the dominant caste. Everyone knew that it was a warning to the Smith to keep to his place, on pain of meeting the same fate as a Bania who had been compelled to leave the village about six years previously. This Bania had been robbed of all his possessions in a nocturnal attack planned by Anavils but executed by their Koli clients. It was characteristic of the Anavils, my informants said, that they had been the first to go to the Bania's house the following morning to express their sympathy. I was told that the Smith, realizing what was in store for him, chose to resist. Three times an attack was announced by letter, and although everyone knew it would not be carried out, the Smith each time informed the police in the *taluka* town. Several village policemen kept watch over his house on the three evenings concerned and, besides food, demanded a kind of hazard money. In this way the Smith paid a penalty of 150 rupees; in addition he made it publicly known that he would stop dealing in mangoes.

Gandevigam has some Anavils, too, who buy up mangoes from the smaller garden owners and sell them independently of the cooperative; they feel threatened by the competition of some Kolis from the village. One Koli, who had bought up the crop of a Moslem grower from another village, complained to me that the Moslem had gone back on his decision; he had been advised by certain Anavils that the Koli was not solvent.

Finally, the ascendancy of the dominant caste is clearly shown by the ineffectiveness of the Tenancy Act of 1957. It is true that ten Kolis managed to obtain some land, but in all those cases the landlord lived outside the village and was not an Anavil Brahman. The Anavils have evidently been able to prevent the transfer of land to their former sharecroppers and tenants. Various Kolis told me that they had not wanted to lay claim to any land to which they had no rights. Their views on unlawful ownership are undoubtedly strengthened by the pressure of the Anavils for whom they are still sharecroppers in secret. A Koli who had himself registered as owner would of course thenceforth not be considered for a sharecropping partnership. It is therefore not surprising that some Koli farmers, most of whom had not enough land of their own to live on, spoke with abhorrence of some fellow caste members who had availed themselves of the opportunity to acquire a piece of land formerly belonging to their landlord.

Nevertheless, as in Chikhligam, the power relationships in the garden village are changing. As I shall explain below, the dominant caste has as yet little to fear from the landless Dublas. Among the Kolis, however, there is a tendency to rise in the social scale, as everywhere in the plain of south Gujarat. This is due to the bettering of the economic position of this middle caste in recent decades. A minority has declined, but most of the Kolis have managed to improve their income. Higher earnings in agriculture are only part of the reason for this improvement. Increasing numbers of Kolis have found work outside agriculture, in which the rise of the level of education in this caste seems to be a factor. Many more of the young people go to secondary school and even to a university. In the process of differentiation that is taking place among the Kolis, those who are most successful try to utilize their stronger economic position to gain political power, by making their weaker fellow caste members feel obligated to them. In several villages the Kolis have already taken over the leadership of the village council,[10] but such shifts of power remain confined to single villages. This largest caste of south Gujarat has not yet succeeded in ousting the Anavil Brahmans from their dominant position in the region, that is, in getting into a position themselves in which they can impose their will on others.

Gandevigam's Kolis are as yet no match at all for the Anavil

10. For extensive documentation on the improvement of the Kolis' position, see Hommes, 1970.

Brahmans of the village, but their more favorable social and economic circumstances make them less dependent than the landless Dublas. This explains why Kolis venture more openly to oppose the landlords of the village and speak bitterly of them. In my conversations with Anavils, their fear of the rise of the Kolis was a recurring theme. They have found that the Kolis no longer behave with docility and are showing a self-esteem which, in the Anavils' eyes, is misplaced. They cite local examples of "faithless" Kolis, and claim that the leasing of land for sharecropping to members of that caste is not without risk.

Yet the dominant caste of the Anavil Brahmans is not really in danger. They act very high-handedly and show uninhibitedly their indifference toward the lower castes. One of the many instances of this is the way in which they behaved over the inclusion of Gandevigam in the electricity network. Electricity was installed only in the streets and houses of the Anavils because, they said, the Kolis could not afford the luxury of electric lighting. This is true of most Kolis: the cost of putting in an electric outlet has to be paid by the occupant of the house. But public lighting, of which the construction and operation costs are financed from the village funds, was confined to the Anavil streets as well, despite a request by the Kolis that some lights be provided in their village quarter. The Kolis are indignant about this. The Dublas are not, because they had never expected to be considered for such facilities in any case.

Many Kolis still rent land from Anavils for sharecropping on an informal basis or are in their debt. For these reasons they owe their landlords support and obedience. The interests of the members of this middle caste are, in other words, divided, which thus far prevents them from combining their strength. They find the path to prosperity and influence barred by the caste of landlords. As one Koli said dejectedly, when he summarized our conversation on this subject, "Where there are Anavils, others cannot get on."

8.4 The Pattern of Agricultural Activities

The distance between Gandevigam and Chikhligam is not, I think, more than 10 miles as the crow flies. The climate in the two villages is therefore very similar. However, owing to the dense vegetation of the garden region the day temperature in Gandevigam fluctuates less. The subdistrict has slightly more rainfall than the neighboring Chikhli *taluka*. Irrigation renders the farmers of the garden region more independent of the quantity and yearly distribution of rain. Gandevigam counts fifty-six wells, which chiefly irrigate the gardens. In 1939 the prince of Baroda had a pumping station built in an adjoining village which, like Gandevigam, is on the Ambika River. From that point the water is carried through irrigation canals to the surrounding villages — in Gandevigam to the *kyari* area. The greater part of the arable land can thus be permanently irrigated. As the soil is highly fertile, crops can be grown more intensively than in Chikhligam. To a far greater extent, therefore, the life of the inhabitants of Gandevigam remains concentrated round agriculture.

TABLE 8.4.1

THE CULTIVATION PATTERN OF GANDEVIGAM IN 1961*

Crop	*Acres*
Fruit	131
Grass	66
Pulses	57
Rice	54.5
Tuberous plants, vegetables, and spices	37.5
Bananas	34
Sugar cane	10
Total	390[1]

* Data obtained from the village records.
[1] Of this total, 42 acres produced two crops in 1961. In addition, 9 acres lay fallow that year.

As in Chikhligam, the soil can be divided into *bagayat*, *kyari*, and *jarayat*. Of the roughly 360 acres of arable land, 145 belong to the first type, whereas the second and third types cover, respectively 105 and 110 acres. The *jarayat* is mainly under grass. In recent years the cattle holdings of the large farmers have sharply decreased. The water for irrigation used to be brought up from a well by animal traction, which required two teams of oxen working in turns, but now motor pumps are used. Despite the decrease of the cattle herds, the grass crop cannot feed the animals of all the farmers, so that many of them have to procure feed from elsewhere.

On the *kyari*, rice is grown during the monsoon, followed by vegetables (potatoes, tomatoes, onions, and cabbage), pulses (especially *val*), and red pepper. These crops alternate with sugar cane or bananas. The *bagayat*, the garden area of the village, is largely under orchards, especially mango trees, but to a lesser extent also *chiku* trees. The cultivation of *chiku* (*sapodilla*) was introduced in south Gujarat in the nineteen-fifties. The *chiku* has a moist tropical habitat and bears fruit three years after planting. Another of its advantages, compared with mango, is the higher yield per tree. In the yards and along the paths in the gardens, finally, there grow the traditional coconut palm, lemon shrub, and jackfruit (*phanas*).

Over the years the agricultural economy of Gandevigam has altered markedly as a result of the changeover to the cultivation of fruit trees. At the beginning of this century, sugar cane was the staple crop in the subdistrict, and until after the Second World War it was the chief source of income for the farmers of Gandevigam. Aside from sugar cane, ginger was grown as a market crop. A specialty of Gandevigam at that period was the cultivation of a spice called *pipar*, used in medicine, which was so profitable that it was sometimes called black gold. The agriculture of the area during the early decades of this century, however, was commercialized to only a slight extent. In the *bagayat* region many diverse crops were grown, as appears from the report of the revised settlement survey of the subdistrict in 1907.[11] That report does not give cultivation data for Gandevigam, but it does include two neighboring villages. There, sugar cane covered about 18 percent of the arable land. In the rest of the area some fifteen other products were grown, for instance cereals (including some species of millet, but especially rice), several pulses, vegetables of various kinds, spices, bananas, oil seeds, and tobacco. The Anavils of Gandevigam told me that these crops were grown primarily for home consumption.

Very gradually the proportion of products leaving the village increased, and this was accompanied by more and more crop specialization. At first the products were the traditional market crops, chiefly sugar cane, but in the past few decades the cane fields were transformed into orchards. The mango was not a new fruit in this region: the trees growing in the yards had long been familiar. The uncultivated mangoes were especially used for home consumption. From an administrative report of 1900 it seems clear that about that time some rich Parsis near Navsari began growing the fruit in orchards.

11. *Revision Settlement Report of the Gandevi Taluka of the Navsari Division*, Baroda, 1907, appendix VII. I tried in vain to obtain a copy of the first survey and settlement report on this subdistrict, dated 1891.

Some country mangoes are also to be sent to the North from the British and Gaekwari groups of bagayat villages round Amalsad. As the fruit trees are grown on the field boundaries, graft mangoes, on account of their stunted growth, are not countenanced in these villages; but the rich Parsis of Navsari have bought up land in Vijalpur and Jalalpor and have converted them into pleasant gardens full of superior mango trees.[12]

It was especially after the nineteen-forties that mango culture became widespread in Gandevigam, but agriculture has remained much more intensive in this garden village than in Chikhligam, as is shown by the agricultural calendar.

The following survey is intended only to convey an impression of the most important activities. It is based on information from a number of farmers and agricultural laborers, verified so far as possible by personal observation. I could not establish in detail the labor required for each of the crops and its distribution over the year, although I was aware that more detailed knowledge on this point would enable me to analyze the labor relationships more thoroughly. The duration of my stay both in Chikhligam and in Gandevigam was not long enough for such a detailed investigation, nor was I sufficiently schooled in agricultural economy.

On the *jarayat* the grass is cut some weeks after the end of the rainy season. During the rest of the year some work must be devoted to cutting down *babul* trees if necessary, removing thorns that have fallen from the trees and might be dangerous for cattle, and mending fences.

The *kyari* is largely under rice during the rainy season, and under pulses or vegetables afterwards, when, as a rule, the ground is plowed and harrowed. Rice and pulses are cultivated in the same way as in Chikhligam. Similarly, the cultivators in Gandevigam have not adopted the so-called Japanese method of rice-growing, which requires more labor and capital but results in a higher yield. For the growing of vegetables and red peppers, seed plots are prepared and fertilized, and small canals are dug to convey the water to the beds. After three or four weeks the young plants are set out in well-fertilized soil, which is irrigated every ten or fifteen days. Weeding is done very intensively both among the plants and along the rows. Three or four months later comes the harvest, after which the ground lies fallow until the rainy season. The culti-

12. *Papers Relating to the Revision Survey Settlement of the Jalalpor Taluka of the Surat Collectorate, Bombay*, 1900, 7.

vation of red pepper requires a great deal of labor. Picking can begin as early as eight weeks after transplantation and may go on until far into the summer.

Sugar cane, formerly the most important product, is now grown only in a small area. The ripening process requires a period of twelve to fourteen months, and of all crops it requires the greatest number of laborers per land unit. Depending on the species and the time of harvesting, sugar cane is usually planted on narrow ridges of earth between the end of October and February. Before this is done, the soil has been plowed, fertilized, and harrowed five or six times. After planting, the fields are irrigated every three or four weeks. Up to four weeks after planting fertilizer may still be added. When two months have passed, the ground between the plant rows is turned. Weeding is very thorough, and the ridges on which the cane has been planted are periodically raised to reinforce them and to keep the roots covered. To prevent impoverishment of the soil, sugar cane is alternated with rice as a monsoon crop and pulses as a winter crop.

The *bagayat* region is mainly devoted to mango orchards. There are not many differences in cultivation practices among the growers or between the two villages, although better farmers use more fertilizer, spray the trees more regularly, and irrigate and plow more intensively.

Chiku trees require more maintenance. The harvest period begins at the end of October and lasts through March. After the monsoon the ground is worked; under the trees, whose branches hang lower than those of mangoes, this is always done with a hoe. Then the ground is harrowed, and some time later it is irrigated and again plowed. During harvest, irrigating is done weekly, for the ground should be continually wet. At the end of the harvest time, and before the monsoon breaks, the orchard is again irrigated, plowed, and fertilized. In Chikhligam several garden owners heap up small mounds of earth around mango trees, but in Gandevigam this is chiefly done under *chiku* trees.

Under the young fruit trees which have as yet few leaves and do not give much shade, cultivation of other crops is possible: bananas, tuberous plants (especially *suran*), and ginger (*adhu* and *haldar*). Several Anavils told me that such cultivation of crops under trees is unprofitable on account of high labor costs. The members of a government experimental farm emphatically denied this and said that the reason for the decline in the cultivation of crops under fruit trees must be sought in the present-day habit of the garden owners *par excellence*, the Anavil Brahmans, of taking life easy.

Bananas, which are also grown on the *kyari*, thrive better on the well-drained loamy soil in the gardens on the *bagayat*. The banana plant needs much water, and to reduce the cost of irrigation it is planted before the rainy season, in the months of April through June. Until the rains come, weekly irrigation is necessary. After a period varying from fourteen to eighteen months the bananas are harvested. The plant, being an annual, must then be taken out. Like sugar cane it is cultivated on ridges that are maintained by raising. As with all wet crops, intensive weeding around the banana plant is necessary. Between the plant rows this is done by means of a harrow, but the weeds among the plants on the ridges are removed by hand. If growth is slow at the beginning of the monsoon, fertilizer is added, being mixed with the upper layer of the soil by harrowing.

The chief tuberous plant cultivated in Gandevigam is elephant's foot (*suran*). Shoots of the ripe tubers are stored in a dark place and are planted on ridges in the gardens about May. Planting in this period enables irrigation cost to be kept low. As with sugar cane and bananas, *suran* is a crop that requires highly intensive labor. Until the rainy season sets in, the fields must be irrigated every ten days. By the end of March or in April, at any rate before planting, the ground is green-manured, after which it is irrigated and plowed. The *suran* takes three years to ripen, by which time the tubers weigh five pounds or more. They are uprooted in January or February and are kept in a dark place until May. *Suran* that is not fully developed is sometimes taken out of the ground and sold in August, when prices for vegetables and tubers are high.

Ginger is cultivated in much the same way. To begin with, the soil is prepared by irrigation, fertilizing, plowing, and preparing ridges. After planting, irrigation is necessary every eight days until the rains come. During this period weeding takes place and the ridges are raised again. The ginger is ripe when the leaves of the plants become yellow and fall off; this is about seven months after planting. The tubers can be left in the ground for some time without spoiling, provided that irrigation is omitted. As is indicated by the foregoing brief outline of the agrarian cycle, in Gandevigam as elsewhere most care is devoted to the cultivation of the market crops. The production of rice, pulses, and grass could undoubtedly be increased and their quality improved, but these crops are grown mainly for home usage. Understandably, it is the market products that receive most attention.

TABLE 8.4.2
AGRICULTURAL CALENDAR OF GANDEVIGAM

Period	*Jarayat*	*Kyari*	*Bagayat*
July through September	After rainfall, cutting green grass	Mixing the soil with water ("puddling"), transplanting rice seedlings; cutting green grass on field embankments; weeding	Cutting green grass in gardens and along paths; banana harvest (planted the previous year); weeding and raising ridges of bananas, *suran*, ginger, etc.; harvesting unripe *suran* and ginger; if necessary, fertilizing orchards
October and November	Making hay some weeks after end of monsoon	Harvesting and threshing rice; preparing soil for winter crops: fertilizing, plowing up, harrowing, digging irrigation canals; sowing *val* and planting vegetables and seedlings of red pepper in seed plots; harvest of sugar cane	Irrigating, plowing up and harrowing orchards; harvest of ginger; beginning of *chiku* harvest
December through March	Cutting down trees, if necessary	Transplanting vegetables and red pepper plants from seed plots; harvest of sugar cane continued, harvest of *val*; irrigating, harrowing, weeding, and raising ridges of sugar cane, vegetables, and red pepper	Harvest of *chiku* continued; irrigating, plowing up, and harrowing mango orchards at first bloom; weekly irrigation of *chiku* trees; harvest of *suran*

TABLE 8.4.2—*Continued*

Period	*Jarayat*	*Kyari*	*Bagayat*
April through June	—	Harvest of vegetables and red pepper continued; mending embankments and fences; plowing up soil and preparing seed plots; at first rain, planting rice seedlings	Mending fences; if necessary, green-manuring with hemp and sugar cane leaves in preparation for cultivating tubers; irrigating, plowing up, and harrowing gardens; planting *suran*, bananas, and ginger; harvest of mangoes, pruning trees

The crop pattern of Gandevigam is more intensive than that of Chikhligam in all respects. Not only is most of the arable land in the garden village under cultivation all year, but the crops themselves require more labor and capital. During a certain period of the year, even more than one crop is grown on parts of the *bagayat*, although this area is rapidly diminishing. Owing to the varied crop pattern there is continuous employment in agriculture, but it declines in the months before and after the rainy season. In April, May, and June, the *bagayat* crops require many hands, while after the rainy season there begins a busy time especially on the *kyari*.

8.5 Landlords and Agricultural Laborers in the History of Gandevigam, and the Changes in the Crop System

Until several decades ago, the relationships between Anavil Brahmans and Dublas in Gandevigam were institutionalized in the *hali* system. As in Chikhligam, I often had to rely on distorted accounts of the way it functioned. Distorted, because the description of a better past reflected the dissatisfaction of both landlords and agricultural laborers with the present. Essentially, however, their accounts were confirmed by what is recorded in colonial administrative reports and older monographs about villages of the plain of south Gujarat.

The Dublas of Gandevigam were employed as *halis* by the landlords of the village. According to my informants there was no category of casual laborers in those days. The Dublas, much less numerous than today, all lived in the gardens and, if only for that reason, all of them had a master. Most members of the caste of agricultural laborers worked throughout the year for their landlord, but when the Anavil was not in a position to maintain his subordinate continuously, the latter was allowed to work for someone else occasionally and keep the earnings. Yet he was regarded as a *hali*, not a casual laborer. In other words, all agricultural laborers were in principle attached; this is the chief difference from the traditional relationships in Chikhligam, where besides *halis* there were also casual laborers. This difference is no doubt due to the dissimilar character of the agrarian economy of the two villages in the past. Because of the very intensive garden culture of Gandevigam, it was advantageous to employ agricultural laborers in permanent service. The need for extra hands was not confined to a few months but was evenly distributed over the seasons. This was the main reason for the Anavils' preference for *halis* over casual laborers — as extra labor, it is true, for the Anavils of Gandevigam used to work the land themselves. Although a few of the dominant caste, the Desais, had acquired various seignorial rights as revenue farmers and occupied an unassailable position as the regional elite, the majority of Anavil Brahmans led a relatively simple existence on the local level.

Within this framework they were without a doubt the most prominent villagers, owners of the majority of the arable land. Yet they were primarily peasants, who distinguished themselves from the Desais by a more unpretentious way of life. Nevertheless the Bhathelas of Gandevigam also kept one or more farm servants if possible. The task of the housewife was then lightened by the maid.

Those who could not afford a farm servant depended in the first place on their own working ability. Of necessity they carried out the physically strenuous and less favored jobs, such as drawing water and looking after the cattle, and they were not even free from plowing. The Bhathelas who employed farm servants were themselves actively concerned in agriculture, and they worked with their servants on the land. At that time, the difference between master and servant must have been apparent chiefly in the kind of work they did.

The Anavils of the garden region certainly knew about the "lordly" way of life of their more illustrious fellow caste members. In the capital of the subdistrict there lived a Desai family regarded as one of the most prominent in the whole of south Gujarat. Undoubtedly individual Anavils in Gandevigam aspired to a life in such grand style. In the village nucleus is to be seen a large three-storey brick house, now in a state of decay. It bears witness to the former wealth of a Bhathela family, whose last son departed in relative poverty to Africa years ago. For a few generations, members of this family had lived extravagantly, like maharajas, some villagers told me with a mixture of admiration and derision. But they overshot their mark, and in the end all their possessions — land, cattle, and also the many *halis* in their employ — had to be given up.

Another Anavil, one of the richest men in the village, told me that his grandfather had become very prosperous in his time. He had owned 100 acres of land and twenty oxen, and employed twelve *halis*. The grandson still took pride in the fact that his father had sworn never to eat without guests. His fellow caste members did not contradict that the oath had been sworn, but they did deny that the father had kept it.

Stories of a glamorous personal past are, however, exceptions. The present generation of Anavils in Gandevigam still remembers the meager existence they led during the crisis years before the Second World War. Many lost their land and had to dismiss their farm servants temporarily or even permanently. But they depict the life of earlier generations in brighter colors, and this agrees with the description of the situation of the Bhathelas in administrative reports of the last century.

> The position of the ordinary bhathela peasant has improved under British rule. Though less frugal than the Kaira Kanbis, they are successful cultivators.[13]

> They in general possess the largest holdings of land. They are active and industrious, and, with the assistance of their Halees or hereditary servants, are enabled to bestow much labour and care on cultivation. They cultivate the major portion of the superior crops and to their exertion and labour the great increase in the growth of sugarcane during the last ten years is to be ascribed. The result of their industry and activity is, that they enjoy a degree of comfort and wealth unknown to the cultivator of the Deccan. The Battellah always lives in a well-built brick house. He is always well-clothed, has the best cattle in the village, and the command of a little money. He is looked upon as a person of influence and substance by others.[14]

13. *Gazetteer of the Bombay Presidency,* vol. IX, Baroda, 6.
14. Bellasis, 1854, 2.

In the eyes of the writer of these reports, the Bhathelas distinguished themselves favorably from the lazy and extravagant Desais by the care they devoted to agriculture.

These are doubtful compliments. It was because of their status and behavior as agriculturalists that the Bhathelas were looked down upon in their caste. The villages in the *bagayat* region, a domain of these peasant Anavils, were regarded as decidedly backward. The functioning of the *hali* system in Gandevigam should be seen in this light. For the majority of Bhathelas, employing servants had a predominantly economic basis, that is to say, it was dictated above all by the very varied and labor-intensive crop system. Extra labor was indispensable in addition to the personal efforts of the members of this landholding caste *par excellence*. In Gandevigam, *halis* were not primarily taken on for reasons of "conspicuous leisure." The patronage character of the *hali* system, although it was certainly not absent, was therefore much less pronounced here than in villages that contained Desai families.

The distance between the *halis* and their Bhathela masters was less marked than that between *halis* and the Desai landlords in other villages. The Bhathelas were less well housed, and they dressed as peasants. A few older Anavil women in the village still wear their *saris* bound up between their legs, a *kaccha* way of dressing which is found only among women of low castes who work on the land. When I asked whether the wives of Anavils had ever been actively concerned in agriculture, my informants always denied it emphatically and indignantly. True or not, this denial was hardly surprising in view of the derogatory proverb that states that the women of this caste used to take their husbands' midday meals to the fields. This in itself is an unheard of and improper activity for Brahman women, let alone agricultural labor. The less refined life of the Bhathelas was also indicated by their food, which differed from that of the other villagers in quantity rather than in quality. This is not to say that the Anavils of Gandevigam approached the subsistence level of the caste of agricultural laborers, or even that of the Kolis. As measured by external criteria of prosperity, however, there was a striking difference between the seignorial style of the Desais and the peasant way of life of the Bhathelas.

The small distance between them and the Dublas was apparent in their daily interaction. The Bhathelas worked on the land with their farm servants and shared their efforts. They were continuously in their vicinity. The Dublas of Gandevigam put great emphasis on the simplicity of the masters of the old days, with whom they worked together, as they claimed, and with whom they were on good terms. Daily contact, not only between master and servant, but also between the members of their families, tended to lead to considerable intimacy. References to a happier past should of course be regarded with caution. Owing to the present antagonistic relationship pattern, both parties are inclined to depict the past situation in rosy colors. Nevertheless, it seems that, for lack of a clear-cut conflict of interests, easy ways and mutual affection were not necessarily exceptional in the relations between the masters and their servants.

The relationships between landowners and agricultural laborers in Gandevigam were altered by the changes of the past few decades in the pattern of agricultural activity. To summarize these changes as the transition from a closed village subsistence economy to an open market economy would be an unwarranted simplification. It is true that in former times most of the produce remained in the village or was directly exchanged for other agrarian and non-agrarian consumer goods in local markets.[15] As has been mentioned, however, everywhere in south Gujarat the growers sold sugar cane and, in the *bagayat* region, other garden products outside the village. Nevertheless, the terms of money economy and market production apply only in part to the traditional agrarian structure of Gandevigam.

Money was used for only a few purposes, especially for paying land tax, for buying land and cattle, and, above all, for financing marriages. It was scarce, and was rarely used by the villagers in everyday life.[16]

It should be added that the closed nature of the village economy

15. "Nowsaree, Gundevee, Chikhlee, Balda, Pardi, and Damaun, and Surat, are all places at which the agricultural produce of these purgannas is largely consumed, and such surplus produce which is not consumed at these market towns is readily bought up for exportation, the facility of water-carriage to other parts being great. Besides these market towns, much business is carried out at the Atwaras, or weekly fairs, held in different villages of the Bhugwara and Parnera purgunnas. Banians from Bulsar, Gundevee, Chikhlee, Balda, Pardi, and Damaun regularly frequent these fairs, with a variety of price goods, cloths, cutlery, cooking utensils, beads, bangles, Native ornaments, pepper, ginger, tobacco, and such articles of general consumption and luxury. The neighbouring villagers attend, bringing with them garden produce, wood, salt, and grain. The mode of transacting business is generally by barter, and little money is used." (Bellasis, 1854, 6.)

16. Joshi, 1966, 38–40.

was breached to only a limited extent by the cultivation of the cash crops referred to above. The contacts of the landowner with the market remained confined to the individual trader from the regional town who visited the villages at harvest time to buy up the produce, either for his own account or on commission for merchants in Bombay or north Gujarat. The sugar cane was pressed by the farmers themselves and from its juice *gur* was made, a compact brown substance that was put into earthenware jars and transported by oxcart. Often the urban trader acted as moneylender, and the peasant became indebted to him when he obtained an advance for one of the spending purposes that have been described. Thus, money transactions were clearly separated from the traditional local form of exchange.

The modernization of agriculture in Gandevigam was gradually completed. As in Chikhligam, it was the result of two reciprocally strengthening processes. First, there was the contraction of the very varied crop system. The wide range of crops, intended to provide as much as possible for the needs of the village itself, was replaced by a much narrower range of produce intended for the market. The extension of the irrigated area in the village was of great importance. Just before the Second World War a distribution network was completed in the *bagayat* region, utilizing the water from the Ambika River. Labor demand was greatly influenced by the fact that the extension of the market production was accompanied by the gradual transition from sugar cane to mangoes as the main money crop.

Secondly, the marketing of this more commercial agriculture was organized along other lines. By the end of the nineteen-thirties a cooperative sugar factory began operating in the subdistrict to process the cane from the surrounding villages. The farmers in Gandevigam became more market-minded: formerly, the factor-moneylender had, as it were, screened them from the market mechanism. Improved road communications with the outside world changed all this. As no permanent crossing facility existed, the farmers used to ford the river early in the morning with their oxcarts and travel the country roads to the railroad town where their garden products were auctioned. On their return they often had to wait for hours on the river bank, because at that point on the river the tides are still perceptible. It was therefore an enormous step forward when a bridge was built just outside the village, and when the road between the two subdistrict towns, which passes Gandevigam, was paved.

The greater accessibility of the *bagayat* region facilitated the founding of an agricultural cooperative with its own selling apparatus. This cooperative, which was initiated by some Anavils from Gandevigam and other garden villages in 1944, has flourished. In 1960, a total turnover of 700,000 rupees was reached. The cooperative is situated on the asphalt road just outside Gandevigam. There is a large shed, in which the produce — especially mangoes, but also *chikus*, bananas, and *suran* — is stored. Operations are conducted by two fully salaried employees, who work in an administrative building equipped with office furniture and a typewriter, as well as a telephone, by local standards the last word in modernity. During the mango harvest, extra hands are hired to unload the oxcarts which bring in the produce, to weigh the baskets under supervision, or to load them directly on to the trucks of the urban merchants. These casual laborers work on piece-rates, and are mostly women from the caste of shepherds (*Ahirs*), who are selected because the Anavil board members say they work much harder than Dublas. The value of the lots is provisionally determined on the basis of the current day's price, which is passed on from the regional auctions by telephone. The amounts are entered on the members' accounts, which are settled at the end of the season.

Early in 1963 the cooperative bought a tractor, which is hired out to members, together with a wage laborer. During my stay the board was planning to install a small factory for canning mangoes. Various negotiations were being conducted, and some members of the board traveled to central Gujarat to see a canning factory in operation.

Improved communications not only have led to commercialization of the marketing organization, but have also increased the contacts of Gandevigam's inhabitants with the outside world, and this was further facilitated when bus services began operating. Owing to the contraction of the agricultural economy, local demand cannot wholly be met by home produce, on the other hand the increasing circulation of money has brought many new articles into the village. Commercialization, in other words, is not confined to production and distribution, but has entered the consumer sphere as well. Villagers regularly go to town to buy what they need.

Owing to these developments, contact with the urban environment has also been intensified.

The commercialization sketched above was partly due to the extension of contacts beyond the village system. Encouraged by the improved means of communication, landowners have allocated an ever larger proportion of their arable land to market production.

In explaining the changeover to mango cultivation, the Anavils, like their fellow caste members in Chikhligam, give as the most important reason the occurrence of plant diseases among the traditional garden crops. They also stress the expectation that the increasing demand for mangoes would render cultivation of these fruits very profitable. The transformation of the garden lands into orchards began before the Second World War, but did not cover large proportions until the end of the nineteen-forties. Low prices for the traditionally cultivated crops cannot have been the reason. Financing the new culture was made possible precisely by the high profits during the war years on such crops as sugar cane and bananas. Referring to that period, one Anavil said to me: "Money came like rain in the monsoon." The garden owners did not seize this golden opportunity to intensify agriculture by using more capital to bring about increased production. On the contrary, they preferred to invest the profits from the war years in such a way that they could adjust their behavior to what they regarded as representative of their caste. Mango cultivation freed them from manual work and enabled them to reduce their active participation in agriculture. This was their most important incentive for changing to the new crop, which does not require intensive labor. As in Chikhligam, the planting of orchards still goes on. Only the least prosperous Anavils have not yet changed their gardens into orchards, partly because they cannot as yet afford the large investment, amounting to about 2,000 rupees per acre. By sinking new wells, the larger landowners have recently reclaimed even the high *kyari* land for mango cultivation. Formerly they grew many different crops under the young trees to offset their temporary loss of income. Nowadays they do not trouble to do it themselves, but lease out the new orchards for this purpose to Kolis. Most of the Bhathelas from the *bagayat* region gradually realized their aim of improving their reputation within the Anavil caste. No longer can they be ridiculed as farmers — toiling drudges who behave like peasants and are no credit to their caste.

Improved communications with the outside world, so closely connected with increasing integration into the market, also contributed to the increased social acceptability of the Bhathelas. In former times an isolated location was looked upon as a measure of backwardness.

The Anavlas at this side of the khadi were formerly loth to give their daughters in marriage to the other side; but now that the danger and inconvenience of the crossing have been removed, the khadi villages do not pay a larger dowry than is customary in other villages.[17]

The Bhathelas from the remote garden villages were at a disadvantage in the marriage market. As they came to lead a more "civilized" life, the pattern of marriage relationships changed. Although I have no specific information — one of the lacunae in my research data — I think I am justified in asserting that phenomena of poverty such as occurred among the older Anavils of Gandevigam (marriage by exchange, bachelor households) are no longer found among the younger generation. Partner selection today is more in accordance with the Anavil ideal. Marriages are contracted with Anavil families from villages which in the past would not have dreamed of entering into any relationship with the Bhathelas of this garden village.

I have attributed the change in the crop system of Gandevigam during the last few decades primarily to the endeavors of the farmer Anavils to attain more esteem within the caste by drastically reducing their participation in physical work. As in Chikhligam, however, the rapid diffusion of mango cultivation brought with it a commercialization of the agricultural economy which had far-reaching consequences for the position of the Dublas in the village.

One of these consequences was the introduction of money wages. After the war ended, the Anavils ceased to pay the wages of their farm servants in kind, and also, as occurred elsewhere in south Gujarat, abolished the various extra allowances that used to be customary.[18] The increasing orientation of the Anavils towards the market brought with it a more businesslike management of their affairs. During that period of high prices for agricultural products it was more advantageous to pay money wages. At the same time, because of the reduced cultivation pattern it was no longer possible to set aside a substantial part of the total produce for redis-

17. *Papers . . . Jalalpor Taluka*, 1900, 5.
18. C. H. Shah, 1952, 448–50.

tribution within the village, for instance as emoluments to farm servants. An ever greater proportion of the agrarian produce was sold in the market. The growing trend toward a money economy in the village depressed the level of existence of the Dublas, for whom the extra allowances in kind had formed, particularly from the qualitative standpoint, an important wage component.

The more general use of money has, moreover, widened the differentiation in ranges of consumer goods used by landowners and agricultural laborers. Because of the enlargement of economic scale, numerous new articles were introduced in Gandevigam, which are exclusively in the possession of members of the dominant caste. They are the only villagers who can afford to buy a great many previously unknown articles — new foodstuffs and luxury goods, such as superior kinds of cereals, canned cooking oil, cigarettes, and tea. There are also the more or less durable consumer goods, varying from cigarette lighters to spectacles, fountain pens, rubber slippers (made in the Bata factories in Bombay), bicycles, large steel cupboards, sofas, stainless steel kitchen utensils, radios, table fans, tape recorders, and other "novelties." It would be incorrect to suggest that previously the difference in housing, clothing, and food between landowners and agricultural laborers in Gandevigam was small, but without a doubt the prosperity gap has widened enormously, both qualitatively and quantitatively. As I have said, the members of the dominant caste have become strongly urban-directed in their habits. More than the other villagers they go by bus or bicycle to the neighboring towns to shop, to arrange business matters in government offices, to go to the barber, to visit tea-stalls or the cinema, in short, to do all those things which notably the Dublas cannot possibly afford.

Production for the market and a money economy, more generally a process of enlargement of scale, have led to increasing differences in prosperity between the landowners and the agricultural laborers of Gandevigam. Gradually the relationship between Anavils and Dublas has changed from a personal into an impersonal one. An increasingly functional evaluation of the Dublas' profitability as labor has been accompanied by progressively smaller remuneration of their services.

Like those in Chikhligam, the garden owners of Gandevigam did not change over to mango cultivation because of fear of labor shortage. If the Anavils now say that at the time their preference for the new crop was partly inspired by their wish not to have to depend any longer on the agricultural laborers of the village, they mean that they simply no longer want to be responsible for them. Epstein has formulated the social background of the patron-client relationship as follows:

In a system in which employers are obliged to provide for the subsistence of their hereditary servants and the latter have no other income earning opportunities, masters will employ labourers as long as the total product does not decrease through the employment of an additional worker. Since the employer would only have to give in charity what he saved in wages, he incurs no net loss by engaging labourers whose employment does not add anything to the total product. The number of workers employed in India's non-market peasant societies is thus determined by the maximization of total product, rather than by the interaction of marginal productivity and wage rates.[19]

This principle, on which the traditional master-servant relationship had functioned, no longer obtained. The familial character of servitude was lost. The contrasts between the two parties sharpened, and neither felt obliged any longer to promote the other's interests. The distance between the two categories, formerly bridged by common aims, became wider, and the *hali* system was bound to disintegrate.

It seems likely that the enlargement of economic scale described above has played a more important role in this disintegration than political and administrative intervention from outside. In Gandevigam, as in Chikhligam, the disappearance of the *hali* system was not due to measures on the part of the authorities, although according to the letter of the law it should have been. Whereas in the part of south Gujarat that was under British rule the *hali* system was never officially prohibited, the principality of Baroda declared it to be against the law as a form of unfree labor in 1923. Nevertheless, it soon became evident that the prohibition had little effect.

In July 1923, the Government of the State by proclamation declared this whole system of forced indenture as illegal and allowed the Raniparaj serf to repudiate it if he chose to. But the intentions of the Government were not properly interpreted by subordinate revenue officials and the operation of the Government's order is therefore not effective.[20]

19. Epstein, n.d., 9; this mimeographed article was published in a slightly different version in 1967.

20. *Census of India*, 1931, vol. XIX, Baroda, pt. 1, 255.

These negative results are not surprising, considering that most of the lower authorities, who had to implement the law in the territory of the principality in south Gujarat, came from the dominant caste, that is, from among the landlords who would be vulnerable if lawsuits followed. As a rule, any conflict between *dhaniamo* and *hali* remained out of reach of the government. If local functionaries were called in, this was usually on the request of a landlord who wanted his rights as a master — not legally sanctioned — to be restored to him. In the existing relationship pattern it was simply unthinkable that a Dubla should address himself to the government with a request for aid. Yet the prohibition did not quite remain without consequences. Its enactment undoubtedly had the psychological effect of making it clear to the Anavils that the government no longer allowed itself to be guided by the Anavils' interests alone. It was an indication of the widening separation of the former power elite from the administrative apparatus in a process of bureaucratization.

Nor did Independence bring immediate improvement in the position of the agricultural laborers in the Gandevigam area. The political and social climate of the principality did not favor the struggle for national freedom, and the lower authorities in particular were unwilling to promote the interests of such underprivileged groups as agricultural laborers. Social workers did not take up the case of the Dublas of the Gandevigam area so long ago that the disintegration of the *hali* system there can be ascribed to their efforts.

With regard to Chikhligam I found that the migratory labor of the Dublas was a result, not the cause, of the changed relationship between landlords and agricultural laborers. In the case of Gandevigam, too, the rise of alternative employment was probably a secondary factor in the disappearance of the traditional service relationship, although after the event such things are difficult to prove. It is true that in the central plain of south Gujarat, notably in the towns along the railroad, a process of economic expansion began even before the Second World War, but agricultural laborers from the villages in the neighborhood were especially at a disadvantage when it came to profiting by the labor demand that accompanied it. As was apparent in Chikhligam, and as will be discussed in further detail below, the Dublas' mobility is subject both geographically and economically to the assent of their traditional employers, the Anavil Brahmans. That is to say, the labor needs of the village ultimately determine whether any Dublas succeed in getting work outside it. In this respect, however, there is a marked difference between the situation in Chikhligam and that in Gandevigam.

To conclude, in both villages it was primarily the landowners who started to change the nature of the service relationship with agricultural labor. In the framework of a process of economic enlargement of scale they began to manage their property in a more businesslike way. One aspect of this change was a pronouncedly impersonal evaluation of the use of those who traditionally served them. Under the influence of their increasing orientation to the market, the Anavil Brahmans of Gandevigam renounced their obligation to provide for the Dublas, and the chance offered to these landowners by the new crop to reduce their own physical activity widened the gap between the two categories.

8.6 The Present Situation

THE DUBLAS. As to the Dublas who live in Gandevigam — 381 in all, distributed among sixty-one huts — my data show that 103 belong to the male working population over sixteen. Of this number, seventy-three work in agriculture and thirty in other occupations. The agrarian Dublas can be divided into farm servants (fifty-one), sharecroppers (seven), and day-laborers (fifteen), but they are only a part of the whole working population of this caste. Most of the female members of the Dubla households also work on the land if there is a demand for their labor, and in addition several of them serve as maids in the houses of the landlords. Secondly, I did not include working children, especially boys, in my census. As in the case of the Dubla women, to include them in the working population would be completely arbitrary. The Dubla boys, who usually work as farm helpers for the Anavils, are no problem, but the question of whether children between, say, eight and fifteen years of age belong to the working population in fact depends on how regularly they are called in. This I could not systematize. Younger children occasionally help out on the land, but they are not remunerated separately. They are not recognized or paid as independent hands until they have reached the age of sixteen.

Taking the foregoing facts into consideration, the involvement of the Dublas of Gandevigam in agriculture is even more marked than may appear from the above figures. The differences in the economic activities of this caste, as between Gandevigam and Chikhligam, are great. Whereas in the latter village 40 percent of the Dublas gain their livelihood by agricultural labor, in the garden village 70 percent of their fellow caste members depend upon agriculture.

FARM SERVANTS. The proportion of farm servants is higher in Gandevigam than in Chikhligam, being 70 percent as against 55 percent of all those engaged in agriculture. The disparity in the composition of the working population within this caste becomes even more apparent if it is taken into account that in Gandevigam almost half of all Dubla men in the productive age group are in permanent service with a landlord. With the exception of one Koli, the latter are all Anavil Brahmans, most of whom also live in Gandevigam.

Servitude is still a relationship that is entered into at an early age, but today it is rarely preceded by a period in which the Dubla boy works as *govalio.* The landlords' stock of cattle has considerably diminished, partly because of the mechanization of irrigation, so that most Anavils no longer employ cowherds. They have found a cheaper solution by jointly appointing a village *govalio,* an old and redundant farm servant. He collects the cattle from the various farms in the mornings and tends the whole herd on the fallow portion of the village *jarayat.*

Even more markedly than before, an advance of money to a Dubla is the beginning of an agreement that nearly always is contracted on both sides for an indeterminate time. Most often a Dubla's servitude is initiated by his marriage. Of the fifty-one farm servants I interviewed, thirty-nine had received marriage money — to the extent of 300 to 400 rupees — from their present master. Their marriage at the age of seventeen or eighteen had been wholly or partly financed by an Anavil landlord, at whose disposal they had put their labor from that moment onwards. The other servants, who had entered service at a later age, had obtained loans for other purposes, such as to pay the expenses of illness in the household. Sometimes the attachment had developed gradually by an accumulation of small advances. What should be emphasized is that, as a rule, servitude begins with a debt.[21]

21. The only cases known to me of servants who did not begin service with a debt are two unmarried Dublas boys who bound themselves to a Koli for two

Even when I did not ask about it, all those concerned told me emphatically that the agreement no longer is valid for a lifetime. Payment of the debt releases the servant from his obligations. Yet the fact that payment is a possibility, recognized by both parties, of terminating the service relationship does not mean that it ends this way in practice. Like all their fellow caste members, the Anavil Brahmans of Gandevigam complain of the faithlessness of the servants they take on. On the other hand, the impression created by the Anavils that running away is usual nowadays was not confirmed by my data: thirty-six of the fifty-one Dubla farm servants have worked for the same master since their marriage. In this respect there is a marked difference from the situation in Chikhligam. In Gandevigam a period of thirty or forty years' service is much less exceptional. It even appears that servitude in half the cases is still hereditary: twenty-seven Dublas are in the service of Anavil families who employed their fathers before them. Least constant are the Dublas from outside the village, for example, the category of sons-in-law (*ghar jamai*) who settled in the village after marriage. Nevertheless, the observer is struck rather by the continuity in the relationship patterns, once they are established, than by a large turnover.

The exaggerated accounts of the Anavils about the Dublas' faithlessness are undoubtedly inspired by their fear of losing their money. But if breach of agreement is the exception rather than the rule, why do the Dublas themselves think of servitude as a temporary matter? Above all because they attempt to deny lasting dependence even if they know better. This is wishful thinking, because repayment of the amount they have received usually turns out to be impossible, and few Dublas in Gandevigam have managed to do it. As a rule, their monetary obligations to the landlord tend to increase in the course of the years.

The doggedness with which the Dublas maintain that the relationship is temporary is due largely to their unwillingness to look upon themselves as permanently attached. For most of them, this is simply self-deception. The lack of employment both in agriculture and in other fields compels them to enter and continue the service which makes their dependence so apparent. Thus the relationship between master and servant has been reduced to a labor agreement based on a condition of debt.

years, in exchange for food, clothing and 60 rupees a year. They will not receive the money until the period has expired.

The nature of the work and the manner of remuneration make it clear that attachment is at present above all a labor relationship. On days when there is enough work, the servants are occupied in the fields or in the gardens of their masters from 7 in the morning until 7 or 8 o'clock at night, except for an interval at midday. In a peak period it may happen that work is prolonged into the evening. Now that the servants no longer live in the gardens, their effort is confined almost exclusively to agricultural labor. The Dublas are no longer in attendance to do all kinds of jobs in and around the Anavils' houses. The master is rarely seen in the village quarter of his subordinate. Only two of the wealthiest landords have a house servant, one of whom is a Dubla bachelor who lives in his master's house. The other is a Koli who, besides acting as major-domo, is charged with the supervision of the agricultural laborers. Some Anavils employ a Dubla boy who performs various jobs and services that were formerly done by the farm servant. On the whole, the work of the farm servant has become more specific: he is responsible for agricultural tasks in the hours appointed for them. Contacts with the master take place chiefly within this framework of time and place. Moreover, today the agreement is restricted to the servant personally, as it is no longer a matter of course that his wife, and in time his children, will be attached to the master's household. It is true that many a Dubla woman works as a maid, but quite often she does so in another household than that of her husband. The diminished demand for cowherds has also been partly responsible for a more individual relationship between master and servant, in which the latter's wife and children are not automatically included. Both the more specific content and the limited character of the servitude indicate that these Dublas are much less incorporated into the landlord's household than were the *halis*.

Accordingly, the remuneration of the servant is based on his work performance and not, as under *halipratha*, on the needs of himself and his family. For each working day the farm servant receives 8 annas, a morning and a midday meal, and tea twice daily. An increasing number of Anavils have now slightly raised the money wage in the harvest period but no longer provide any food. The farm servant is annually entitled to some pieces of clothing, a pair of shoes, and a shawl, and to remuneration for cutting grass and looking after the cattle during the monsoon. He should be allowed to gather wood in the master's garden, and at the beginning of the monsoon the landlord should provide him with thatch to mend his roof. A little tobacco should be a regular wage component, and in the event of illness the master is expected to pay for the medicines. At harvest time, finally, the Dubla should be able to count on a small share of each crop.

This is how the relationship should be if custom were adhered to. However, the Dublas complain that they are given worn clothes and old shoes, sometimes bought for the purpose in the neighboring town. Medicines either are not paid for by the landlord at all or their cost is added to the existing debt. Prolonged illness has become grounds for dismissal.

Aside from skimping, which became systematic as the money economy became more prevalent, the replacement of part of the wages in kind by payment in cash has been to the disadvantage of the servant. What the Dublas used to be paid in produce they now have to buy at the higher retail prices. The effect of the unprofitable terms of exchange is a curtailment of their consumption.

On days when there is no work the farm servant cannot claim *khavati*. The master expects him first to try to find work with another farmer. Only in a period of continuing and complete unemployment, notably during the monsoon, does the master feel obliged to yield to the Dubla's repeated pleading and pay him at least a minimal allowance. The reluctance of the Anavils to extend this form of credit is due, to no small extent, to the fact that advances, from marriage money to *khavati*, are nowadays given wholly in money. The amount that the Anavils are forced to spend in this way deprives them of it for a number of alternative spending purposes to which they accord higher priority. For the same reason the landlords are not prepared to comply with the requests of their servants to defray high incidental costs — for instance, expenses incurred in cases of illness, a marriage, or a death. The compensation paid to a servant is determined by the work he delivers or will deliver. The time is past when compensation was based on the master's moral obligation to maintain his *halis*.

A continuous need of labor explains in part why the landlords of Gandevigam still take Dublas into permanent service. The Anavils are guided in this not only by the nature of the crop system — which even now is labor-intensive during a large part of the year — but also by their desire to abstain from physical agricultural work. Employing one or two servants means having a cheap, fixed nucleus of labor and a claim to temporary extra hands, for although the agreement covers the servant alone, the latter is expected to mobilize his relatives for the benefit of his master when they are needed. Thus, while reducing his responsibility the landlord can enlarge and contract his stock of labor at short notice. He of course runs the risk of losing the amount of his investment if the Dubla leaves him without notice. He cannot appeal to the government, for the lending of money is now subject to a system of permits. There is even an explicit prohibition against the discharge of debts by means of work. Although a Dubla knows better than to adduce this argument even if he should be aware of its existence, it is of course not without consequences that the claim of the landlord is not recognized as legally valid.

The Anavils arm themselves against the risk of desertion by their laborers by not letting the debt become too high. Restriction of credit is not, however, their only guarantee. Since the Dublas have ceased to live in the gardens, various landlords have proceeded to farm out cattle for sharecropping, as it were. A newborn calf is brought to the hut of the servant, who undertakes to look after the animal, but he can do this only if he has received permission to cut grass in his master's garden. When the cow is in milk it goes back to the owner, and the servant is entitled to half the estimated value of the animal. A variant is that an agricultural laborer receives half the calves bred from a cow that he has looked after. In Chikhligam, where most of the Dublas are absent during a large part of the year, this "lending" of cattle does not occur, nor are all the landlords in Gandevigam prepared to do it. Sometimes the care given the animals by the Dublas leaves much to be desired. On the other hand, this is an excellent means of binding the farm servant who no longer lives on his master's land. It is especially desirable for the Anavils, because in this way they earn back part of their investment and can go on giving credit, while the debt of their subordinates remains within reasonable limits.

The immediate reason why a Dubla becomes a farm servant is usually that for most members of this caste this is the only way to finance their marriage. Yet the loan is concomitance rather than cause. For the great majority of Dublas who perform agricultural labor, servitude still offers the best guarantee of a living. Although their daily wage is extremely low, they are assured of employment on most days of the year. The relation between master and servant has become markedly businesslike. The Anavils have unilaterally fixed a standard wage, above which they need not go. The masters are generally not susceptible to any appeals by the servants for further assistance. Anavils told me that they give their Dublas something extra from time to time when they are in a good temper. The reaction of their subordinates indicated that a good-tempered landlord must indeed be a rare figure in Gandevigam.

On the other hand, there is no longer any question of many-sided service by the family of the Dubla. The Dubla's contribution is personal and as a rule confined to agricultural labor. The allowance he receives is not sufficient for him to be able to cope with eventualities such as divorce and remarriage, illness, continuous rain during the monsoon, or collapse of a wall of his hut. On such occasions the master cannot refuse too often to come to his assistance — at least if he wishes to keep his servant. But what formerly was regarded as part of the maintenance wage is now booked as debt. However, the farm servant has at least a possible source of credit, even though laboriously obtained. Aside from the security of a minimal income this is for the Dubla the compensatory side of servitude. The drawback, however, is profound dependence, for the master manipulates the debt as a means to compel obedience, and sometimes even as an excuse for not fulfilling his minimal obligations. He knows how to demonstrate his authority in numerous ways. Servants are not given the perquisites to which they are entitled, nor do they get *khavati* unless after prolonged insistence. If it does not suit him to do so, the Anavil refuses to pay the wage and tells his subordinate to return the next day. Should the servant be ill, the master comes to his hut only to ask impatiently when he will be fit for work again. An "insolent" Dubla is sometimes not given any work for days on end to teach him a lesson. It is in an attempt to escape from the tutelage and subservience imposed upon him that the farm servant tries to pay off the amount he received for his marriage. My data show that only eight Dublas in the village have succeeded in doing so, owing to a combination of advantages. One had extra earnings as a drum-player at marriages; another worked with his master's permission in a factory for two years and gave him his savings on his return. The others reared cattle for their masters for some years and paid off their debts in this way. All of them pointed to the absence of setbacks over longer periods in their career as servants, such as long illness, having to care for aged parents, death of calves reared, or many small children — that is to say, the absence of circumstances that are normal rather than exceptional for the situation in which the Dublas find themselves.

The example of the fortunate ones is what the Dublas have in view when they say that the relationship is temporary. Yet it would be incorrect to infer that their desire to pay back the marriage money indicates an effort to break the servant relationship. Without a doubt, the large majority wishes, for lack of a better alternative, to remain in servitude. For them, repayment of debt is first of all an attempt to render themselves less dependent on their masters, to lessen their control. Of the eight Dublas who paid back their marriage money, three are still employed as servants, though no longer bound by debt.[22] Four Dublas who were freed from debt servitude through no action of their own are in a similar situation. They became redundant when their former master sold or lost his land, left the village, or died. These unindebted servants now receive from their new masters a slightly higher daily wage, which has only heightened the desire of their less fortunate fellows for this form of service.

It is precisely for these reasons that the Anavils appear to be less than happy with this kind of relationship. They are therefore not at all sorry that so few farm servants manage to repay their debts in the end. For that matter, the landlords have it in their own power to prevent repayment, and they do not shrink from wrongfully adding all kinds of expenses to the debt. The Dublas are well aware of this trickery, but they are powerless. A striking instance is that of the master who went to the hut of a Dubla servant to lecture him about his absence. On the way there he hurt his foot on a thorn, with the result that the Dubla's account was debited with 5 rupees, the amount the master had spent on medicines. Another Dubla entered into marriage at the same time as his brother, and together they obtained 275 rupees from an Anavil. When, after three months, one of them died, the master charged the remaining brother with the total amount.

It is not to be wondered at that in most cases the debt does not decrease but becomes larger. It was my impression that most of the farm servants have long since abandoned hope of ever becoming

22. Of the others, one has found work outside agriculture, and four have become day-laborers.

free of debt. They keep up appearances for themselves and in the presence of others, but at the same time they try to persuade their masters to give them as much credit as possible. Once attached they have little to lose. It is therefore understandable that severance of the relationship is usually due to a conflict between master and servant which has flared up suddenly. This does not happen as often as the landlords claim it does, but no doubt more often than termination by mutual consent after repayment of the debt.[23]

To break with the landlord means in many cases to break with the village and to give up shelter on a piece of land of one's own, at least temporarily, without having any clear notion of how or where to find other employment. A Dubla will think twice before he deserts his master. Often this also means a break with his family. The deserted wife returns to her native village with the children and enters the household of her parents or a brother. Economic hardship of this nature is no doubt the chief explanation of the relatively high frequency of broken families in the quarter of the landless laborers. There are a considerable number of children who, for this reason, are taken in by their mother's brother or mother's sister. Several Dublas can tell me only one thing about their father — that he ran away — and they do not know where he went nor whether he is still alive.[24]

Running away is for a Dubla an act of desperation rather than an expression of indifference and unreliability, as the Anavils claim. The master's reluctance to give credit is, in other words, inspired by an abundance of labor supply, and not primarily by the Dublas' faithlessness. An illuminating remark was made by an Anavil who told me that a farm servant leaves only when his master does not fulfil his minimal obligations. A resolute and repeated refusal to give credit, for instance in the form of *khavati*, means in fact that the advantages of servitude are lost and the disadvantages of attachment remain. Both parties reckon with the possibility of termination of the agreement, but their views as to how it should end differ. The diminished stability of the relationship pattern is the result

23. Five Dublas in Gandevigam admit that they ran away from their masters after a quarrel. Actually there were more who did so, but their number can no longer be ascertained, as several left the village. Conversely, more than one of the numerous sons-in-law in the Dubla quarter came from elsewhere for this reason.

24. In eleven huts, one of the parents of the children living there is absent for another reason than employment of the father outside the village; nineteen children, born outside Gandevigam, grow up without their parents in households of relatives in the village. These are incomplete but telling data.

of the landlords' indifference and denial of total responsibility rather than of emancipation of the farm servants.

SHARECROPPERS. It is open to question whether the sharecroppers among the Dublas, whom I have placed in a separate category, should not rather be included among the farm servants. In giving a plot of land to Dublas for banana cultivation, the Anavils have found a new means of assuring themselves of dependable labor. Only reliable agricultural laborers are considered for sharecropping. Part of their time is at the landlord's disposal, and often a son of the sharecropper works as a farm servant. Apart from this it is profitable for the Anavils to have Dublas, because they are satisfied with a smaller proportion of the produce than Kolis. Finally, there is virtually no chance that the agricultural laborer will have himself registered as owner under the new law. For all these reasons the number of sharecropping farm servants will probably increase rapidly in Gandevigam.

Harper mentions such an agreement between landlords and agricultural laborers elsewhere in India:

> In more recent years a few Haviks have leased land to their servants to induce them to continue in service. . . . Had it not been for the Tenancy Act in 1947 this method of enforcing the system of indentureship might have succeeded. The Tenancy Act gave the servant permanent rights to the land and, in addition, limited the amount of rent that the landlord could ask.[25]

In view of the relations between Anavil Brahmans and Dublas in south Gujarat this last remark seems somewhat unrealistic.

DAY-LABORERS. During my stay in Gandevigam I counted fifteen day-laborers among the Dublas. This is a moderate number, but it should be borne in mind that, according to my informants, some decades ago no one at all in this category was to be found in the garden village.

Although they work mainly for the local landlords, the day-laborers are less bound to Gandevigam and the Anavils than are the farm servants. They look for casual work in the other villages as well, and are called in from time to time by Kolis in the neighborhood. As a rule they receive 12 annas and two meals a day from the farmer for whom they work. At harvest time they are not

25. Harper, 1968, 58.

usually given meals, but the daily wage is raised to 1 rupee or higher. They also occasionally undertake to perform a specific task for a prearranged lump sum of money. Sometimes they are bilked of part of the agreed remuneration for the work assigned to them. When, for instance, an amount of 10 rupees was arranged, the landlord may decide after completion of the job that 6 rupees is really more than enough. The powerlessness of the agricultural laborers in the face of such trickery does nothing to diminish their rancor.

Some Dublas in this category have always been day-laborers. A few are unmarried boys, and for that reason as yet unattached. Others are former farm servants who paid off their debts or were dismissed. Especially among the sons-in-law in the Dubla quarter there are several who have escaped from the control of their former masters and now live as free laborers in their wives' villages. The background of this category is therefore varied. The group is not in the least closed; in several cases I found farm servants and day-laborers in one household. It is difficult to ascertain which of the two is better, or rather less badly, off — the day-laborer or the farm servant. Some of the former much prefer to remain unattached. Being self-assured and resourceful, they succeed in scraping together a day's wage even in periods of continuing unemployment. They belong to the type that can always look after itself and does not wish to be subjected to a master. Others prefer the security of a minimal existence to the risks of an unattached life. The offer of a higher average daily wage is for them no compensation for the possibility of receiving no allowance for days on end. Their employment as day-laborers is often only temporary.

There is indeed no clear-cut difference in conditions of employment. In cases of continuing unemployment or more incidental reverses, the day-laborers try to procure an advance from a landlord. In exchange they undertake to work for their creditor on such days as he may need them and against the remuneration of a farm servant. They are then merely unattached in name. If the debt runs up they become in fact part-time servants, a status which either continues or finally changes into complete servitude according as the Anavil thinks fit.

EMPLOYMENT OUTSIDE AGRICULTURE. Thirty Dublas in Gandevigam work outside agriculture. This represents nearly 30 percent of the male working population of that caste. The rise of their proportion in the past few decades reflects the expansion of alternative employment in the neighboring towns and the improvement of communications. In subsection 8.2 I have given a survey of their activities.

I did not get the impression that these Dublas isolate themselves from the agricultural laborers in their caste and form a separate category, nor did I find any concentration within a limited number of households. Those who have managed to find employment outside the village, however, do try to assist their relatives to find such jobs as well. Owing to their greater familiarity with the urban environment and the means of gaining a livelihood there, they are in a somewhat better position to do this than those who work in the village, although they can rarely exert any influence in the matter. Recommendation by the landlord usually carries more weight. Several Dublas therefore owe it to the support of an Anavil — their father's master, for example — that they found employment in a workshop or obtained a job with some government agency. The Dublas recognize the influence of the landlords in this respect, but are convinced that it is mainly used negatively, that is, to prevent them from finding employment outside the village. They complain that "the Anavils speak badly of us and see to it that we are not accepted." Local government functionaries admitted to me that these charges contain a germ of truth. In their view the landlords are afraid that if alternative employment expands they will not have enough labor themselves, or at least that their authority will be impaired and the wage level will rise.

The twenty-five Dublas who leave the village early in the morning and return in the evening are without a doubt the most fortunate members of their caste. In the first place they have a fixed income, with a higher daily wage than that of the agricultural laborers. Secondly, they have escaped the control of the landlords in the village. Their good luck is relative, however. In the workshops and government offices they receive the lowest pay; most of them do not earn more than 1.50 rupees a day. Those who work in Bilimora, seven miles away, cannot afford to go there and back by bus, which costs 0.40 rupees each way. They, too, have to incur debts in order to live. One Dubla, who works in a paper factory, and two others, who are employed by a wood merchant, have been given advances by their employers to marry. In other words, their independence has become limited.

Of all the Dublas in Gandevigam, only a few seem to migrate to the brickyards every year. During my stay they numbered not more than five. Unfamiliarity with the channels leading to work in the environs of Bombay is only a small factor. Most of them know quite well to whom they should apply, but say they do not feel attracted to this kind of work. The difference in agrarian employment between Gandevigam and Chikhligam, where during the greater part of the year there is little or nothing to do for agricultural laborers, no doubt explains why so few from the garden village avail themselves of this opportunity. It may be that intervention by the landlords is also a factor. Nevertheless, in the existing situation migration is evidently not a desirable alternative to agricultural labor. The Dublas talk at great length about the disadvantages — the strenuous work, the insecurity of the new surroundings, the temporary character of the employment, the small difference in wages, and the uncertainty inherent in migratory labor. With some exaggeration it might be said that the push factor is lacking and the pull factor is not strong enough.

A GENERAL SURVEY OF THE CONDITION OF THE DUBLAS. From the foregoing analysis, it seems clear that the Dublas who work outside the village are least badly off. Some possess a few more

TABLE 8.6.1

NUMBER OF OCCUPANTS PER HUT AND THE DISTRIBUTION OF THE MALE WORKING POPULATIONS OF THE DUBLAS IN GANDEVIGAM

Number of Occupants per Hut	*Number of Huts*	*Number of Occupants*	*Male Working Population*		
			Average per Hut	*Totals*	*Agricultural Laborers*
4 or less	18	56	0.8	15	9
5 through 7	27	168	1.7	45	32
8 and 9	11	94	2.2	24	13
10 or more	5	62	3.6	18	11
Totals	61	380	1.7	102	65

material goods and live in slightly better huts than those of their agricultural fellow caste members. In general, however, the various categories of Dublas do not live in separate households. This means that differences in the degree of poverty are determined primarily by the ratio between productive and unproductive members of each household and only secondarily by the nature of their work.

The Dubla quarter presents a cheerless sight. It is on rough and uneven terrain that is waterlogged during the monsoon. The huts, many of which are in a state of decay, were constructed in a rambling order. A dwelling consists of a single room which one must stoop to enter, so as to avoid the low, overhanging roof. The inhabitants crowd together in a space of a few square yards, in continuous twilight and, except for the doorway, without ventilation. To say that this interior is scantily furnished is to put it mildly. In a great many cases even the simplest household goods for daily use are lacking. I stopped trying to record the goods they did own, because my interpreter found it too painful to go on enumerating so many articles which obviously were not there. Most of the inhabitants sleep on jute sacks or home-made reed mats. Kitchen utensils are usually limited to some earthenware pots and a few copper pans and mugs. An official report on the condition of the *halis* prepared shortly after the Second World War stated that less than half the Dubla households owned a grinding stone. Further, one-third of the 285 families that were interviewed did not possess even the simplest ornaments, which are usually acquired at marriage; in other families their worth did not exceed 5 rupees.[26]

As regards Gandevigam, the condition of the Dublas has not improved in any way since the war. There is so little floor space that not all the members of the household can sleep inside. The furnishing is so meager that, in my eagerness to record at least something, I found myself counting the nails in the beams, on which here and there some clothes were hanging — rags to me, but to the owner a precious second garment. Only the men wear shoes, which are indispensable for work in the field. The children in particular are scantily dressed. As for food, cooking oil is rarely used. Salt and spices are bought in the village shop, sometimes in exchange for the eggs of the hens they keep. Only farm servants and maids regularly eat vegetables as a component of the meals they are given in the household of the master. During the monsoon, when the income of the agricultural laborer drops even further, they are reduced to living on a gruel of water and rice (*barka*), occasionally supplemented by a species of weed as a vegetable. The Koli shopkeeper in the Dubla quarter sells many articles in smaller quantities than the normal weights: a pinch of tea, a handful of chillies, a small bag of salt, etc. Most households are in his debt to an amount

26. *Report of the Hali Labour Enquiry Committee*, 40.

of 3 or 4 rupees, and the shopkeeper does not allow them credit beyond that amount. Accounts are settled when the wages rise at harvest time.

The miserable living conditions and the undernourishment result, of course, in bad health. More or less chronic illness is a familiar and accepted phenomenon in every household. A widespread high consumption of home-distilled alcohol completes the picture of pauperism. Equally typical of the situation is the view of the landlords that alcoholism is the cause of the Dublas' poverty instead of an escape from daily misery. In the 1950 report cited above, the term "subhuman standard of living" was used, a description as valid today as when it was written.

The Dublas have seen the Anavils' prosperity rise rapidly in the last few years and compare their own deplorable conditions with the landlords' comfortable way of life. Most of them feel that their situation has deteriorated within the past few decades. It is true that formerly they lived in the gardens of the Anavils and were attached for life, but the masters shared the effort and remunerated them more liberally in those days. They were part of the landlord's household and their living was secure. The daily allowance was paid in grain, and the prices of foodstuffs and other articles that had to be bought were much lower. The picture they gave me of the earlier situation is certainly too optimistic, for it took no account of arbitrariness and compulsion. On the other hand, a more miserable existence than that led today by, especially, the agricultural laborers among the Dublas is inconceivable. Aged farm servants who can no longer work are dismissed without ceremony. They can only fall back on the support of any children or relatives who are able or prepared to give it, but these find it difficult enough even to keep their own heads above water. In such a situation of extreme scarcity, mutual assistance even among members of the family is weakly developed, and the prevailing hardness between near relatives accentuates the hopelessness of existence.

The low standard of living of the Dublas in Gandevigam is closely connected with the disproportionate growth of their numbers. All informants pointed to the high increase in the number of Dubla huts since the previous generation. So far as I could ascertain, this demographic development is due chiefly to a migration surplus and falling death rates, the latter factor being undoubtedly the more important. Famines and epidemics, to which many formerly fell victim everywhere in India, especially among the castes of agricultural laborers, have become fairly exceptional. The expansion of agrarian employment in Gandevigam has not kept pace with the rapid rise in the number of landless laborers. On the contrary, the crop system has become less labor-intensive, and the changeover to pump irrigation in the gardens has reinforced that tendency. Nor has it been counteracted by the fact that the Anavils themselves no longer work with their servants, and that few Kolis are now prepared to hire themselves out as day-laborers for 8 or 12 annas.

Diminished demand and increased supply of labor has contributed greatly to the weakening of the position of the agricultural laborers. One of the consequences has been the rise of a category of day-laborers. This labor reserve has lessened the necessity for Anavils to safeguard themselves against shortage. The number of farm servants in the employ of landlords is no longer adjusted to the maximal demand of labor at a few peaks of the agrarian cycle. Now that the risk of temporary scarcity has lessened, the investment and maintenance costs of a farm servant, converted into money as they are, are weighed against other priorities.

Yet attachment is by no means disappearing. The distribution of the work throughout the year is such that the Anavils still find it desirable to maintain a permanent nucleus of labor, the more so since most of them no longer work the land themselves. Appointing one or more Dublas in permanent service is, in the long run, easier and cheaper than having to call in day-laborers every time.

Nor have the Dublas found employment outside agriculture for their growing numbers. As I said, the landlords oppose it, because a cheap and abundant labor reserve is more to their interest. Apart from that, alternative employment is hard to get, because of the low level of schooling among the members of this caste. They are considered only for the humblest and lowest-paid jobs, and this will be true for a long time to come. Only a handful of Dubla children attend the village school fairly regularly and for some consecutive years, even though education is compulsory. The school has registered a larger number, but checks and inquiries made it clear that most of the children are rarely if ever present. Whereas the Kolis avail themselves of education as an avenue for strengthening their social and economic position within and outside the village community, most of the Dublas are not much concerned about the schooling of their children. The indifference of the older generation, however, is not the chief reason for the prevailing illiteracy among the younger generation. In the first place, older children must stay home to look after their younger brothers and sisters when both parents are working. Secondly, a number of Dubla boys, who formerly tended the cattle of the Anavils, are now *govalia* for their own father. They are charged with the care of the cow or buffalo which the landlord has farmed out with their father. Finally, the social climate is scarcely conducive to regular attendance of a large number of Dubla children.

Settlement of the Dublas in a village quarter of their own has somewhat lessened their dependence on the landlord. This applies primarily to those who work outside Gandevigam. They are the ones who most strongly resist the Anavil yoke under which their fellow caste members are laboring and who say that they themselves would refuse to work for 8 annas. Yet they fear being driven back into agriculture and thus, finally, into their former subjugation. On account of their vulnerability they dare not risk an open conflict with the power elite. Nevertheless, behind the scenes it is they who fan the discontent that prevails in their caste, who express it, and whose voice is persuasive.

Among the Dublas, the awareness of their subjugation has grown. They are beginning to feel that they should shake it off and cease to concede the superiority of the landlords. Yet the weak economic position in which they find themselves perpetuates their dependence and perhaps aggravates the subjugation by its depersonalized character.

The situation of the other Dublas does not fundamentally differ from that of the farm servants. Although there is some differentiation among the caste members, it rests mainly on differing extents of poverty and attachment, on different forms of dependence.

THE ANAVIL BRAHMANS. Within the past few decades, the Anavil Brahmans of Gandevigam have become landlords. This term does not refer to the quantity of land they possess — which has probably not changed very much, taking the caste as a whole — but more particularly to the degree to which they themselves are physically active on the land. The drudges of former times have developed into gentleman farmers who rarely if ever work the land. It should be added that this does not apply to all the members of the dominant caste. Of the twenty-six Anavil households in the village, seventeen employ a total of forty-three servants, and the remaining nine households have to get along without laborers in permanent service. Three of these landowners do not work at all. They subsist on the proceeds of their neglected gardens or small orchards, and leave the annual picking to a factor who has bought the crop long in advance. Their meager income is supplemented by the incidental allowances from relatives outside the village.

The other six households own too little land to be able to afford the convenience of a servant. Daily work in the fields and gardens is done by the male members of these households. Only when they cannot cope do they call in Dublas or Kolis for the least-favored jobs, notably plowing. Usually such temporary servants are day-laborers, but sometimes they are farm servants who have been given a few days' leave by their masters. The Anavils of this group are hard-working farmers who still grow the traditional garden crops, partly because they are more profitable than fruit. They strive to free their sons from agricultural labor when the time comes. Their poverty is apparent not only from the absence of permanent help on their land, but also from such evidence as a shabby house or the bachelor life of one or more brothers. They remain in the background, and some of them live more or less hidden, deep in the gardens. Although the older Anavils made no secret of their presence they told me with some emphasis that the way of life of these simple farmers was not what might be expected of Anavils.

It is an existence that most of their fellow caste members have happily left behind them. They have attained prosperity and surrounded themselves with a comfort designed to demonstrate their living as landlords. For miles around, the Anavil Brahmans of the garden region enjoy the reputation of possessing two houses, an unbelievable luxury amidst poverty and scarcity. The largest and most recently built garden houses cost from 15,000 to 30,000 rupees, and the wealth of their owners is also apparent from the way the houses are furnished. In contrast to their sober dress of former years the most prominent Anavils now wear immaculate white *dhotis;* they no longer have stubbly beards but are clean-shaven, thanks to their electric shavers.

By avoiding agricultural labor themselves, the members of the dominant caste have accentuated the social function of servitude. Light work is allowed, such as sheaving or mango-sorting, but the

Anavils apply themselves chiefly to the organizational aspects of their undertaking. They distribute the labor tasks and supervise the progress of the various activities. Their old people tell "when I was young" stories and grumble about the boundless laziness of the young. They still have memories of the privations of the nineteen-thirties, and reproach their sons for playing cricket rather than helping in the garden. Nevertheless the older generation, too, accepts with visible pleasure both the newly acquired wealth and the convenient way of life it has brought. Their social need for servants has increased to such an extent that, in spite of the changeover to mango cultivation, which requires less intensive labor, it is doubtful whether the number of agricultural laborers in permanent service has really diminished. Perhaps it is only in a comparative sense that servitude has become less important. For the Anavils, the cultivation of fruit trees has satisfactorily reconciled their two divergent requirements: the highest possible income and the least possible effort. They look with disfavor upon some Kolis who have followed their example on a small scale, but when a Dubla planted a small tree in front of his own hut in the new quarter, this was regarded as intolerable, so some Anavil boys pulled it up.

Between the Anavils of Chikhligam and those of Gandevigam there is one great difference. In the former village the members of the dominant caste have removed themselves, as a rule, much farther from agriculture. Many of them are employed in urban professions, a development that began even in the previous generation. For the Anavils of the garden village, on the other hand, land property has remained their chief source of income, hence they are considerably more active in it. Such complete idleness as in the *jarayat* village is not often encountered here. The landlords of Gandevigam are much more frequently in their gardens or fields, where they supervise the work of their subordinates. Their greater commitment to agriculture is apparent, for instance, in the recent introduction of *chiku* cultivation. This fruit tree requires more care (irrigation and labor) than mango, which for their fellow caste members in Chikhligam is a reason for not taking it up despite the attractive market prices. The Anavils of the *bagayat* region try to avoid activities that would label them as working farmers, but they have no objection to directing a farming enterprise. This means, for instance, attending with great concentration to the daily schedule, watching the cost sharply, and continually checking the progress of the undertakings. In other words, they lay stress on their managerial role.

I have described in detail the consequences of this increasingly depersonalized way of management for the master's relationship to his agricultural laborers. It should be noted, too, that the relations among the Anavils themselves tend to become more businesslike as well. I was told more than once that the old hospitality that marked their social intercourse has disappeared. They themselves explain their preference for living in their gardens by their wish to remain secluded as much as possible, or at any rate to limit contact to what is strictly necessary. The cordial and spontaneous aid that formerly would have characterized their mutual relations has been replaced by an unmistakable aloofness. The members of the dominent caste are always prepared to vilify each other, and with no hesitation they accuse their neighbors of greed or denounce them as liars. When still in Chikhligam I had been warned that the Anavils in Gandevigam "do not collaborate well with each other." Jealousy and spite aroused by newly acquired wealth or esteem lead to all kinds of personal and family quarrels — about land, about laborers, about taking unripe fruits to the cooperative, and so on. The pejorative stereotypes current among outsiders about the dominant caste are also used by its members, but only in relation to other households than their own.

Nearly all of the Anavils of Gandevigam have retained their interest in agriculture, even though most of them no longer work the land. Notwithstanding the vicinity of some urban centers and the excellent communications, only a few of them are employed outside the village. Their fellow caste members in more remote and isolated Chikhligam are much more urban-oriented. What is the cause of this difference?

In the first place, the Anavils of the *bagayat* region belong to the section of the Bhathelas who have always led a peasant existence. In villages where the dominant caste is represented by Desais there was an early tendency towards separation from agriculture. Education took root and developed much earlier in the *jarayat* villages precisely for this reason. A school education facilitated escape from the agrarian environment and was a condition for obtaining one of the coveted positions in the rapidly growing framework of local government.

Because they are still so much concerned with agriculture, the

Anavil Brahmans of the garden villages are held in lower esteem within the caste than are those of the *jarayat* villages. In Chikhligam I was given to understand that I should not go to Gandevigam if I wished to know how Anavils live. Gandevigam was portrayed as a real peasant village populated by Bhathelas.

The urban orientation of the dominant caste has been greatly influenced by the Desai behavior pattern, and the Anavils of the *bagayat* region showed, as it were, a delayed reaction. The time lag, however, is not the only reason why they are as yet so rarely employed in nonagrarian occupations. The profitability of agriculture during the past few decades has undoubtedly been another factor. On the basis of categories which I supplied, three Anavil informants classified their fellow caste members in the village as follows: eight rich; seven well-to-do; seven moderately well off; four poor. Only for the first category had I mentioned a demarcation line, an annual income exceeding 10,000 rupees. In the year preceding my stay the two largest garden owners had earned 20,000 rupees by the produce of their fruit trees alone. The prices of *bagayat* plots have soared, but land is rarely on the market. An Anavil who had been classed as moderately well-to-do told me that he would not sell his garden even for 50,000 rupees. The owners cling to their land because it enables them to have a high income, not simply through pride of possession.

Despite the Bhathelas' adjustment to the style of life of the Desais, their status within the caste has not greatly improved. As mentioned earlier, a number of the traditional standards of prestige have lost significance. A landlord's way of life was the model for the previous generation, but now the general preference, as in Chikhligam, is for residence in town and an existence outside agriculture. It is clear that the Anavils of Gandevigam do not conform to this standard. The high income they draw from agriculture can only partly modify the negative evaluation of their rural way of life. They are now endeavoring to follow the example of their more "advanced" fellow caste members in their orientation towards the town. It was therefore to be expected that I found among the younger generation the same contempt for agriculture and village life as elsewhere. "For farming you need no brains; it is dull and stupid work" was the way in which the educated son of a landowner summarized the attitude of his age group.

After secondary school, many Anavil boys, and increasing numbers of girls, go to college in the neighboring town to try for a B.A. or a B.Sc., the threshold to the middle cadre of government and industry. The richest landlord of the village plans to send his son to the U.S.A. for a year to pursue his studies there, which will cost him about 15,000 rupees. After his return, the son will be able to ask a higher dowry and be in line for a relatively well-paid job in town.

The growth of the number of Anavil Brahmans, though not nearly so rapid as that of the lower castes, stimulates departure from the village. By enabling one or more of his sons to take up employment outside agriculture, a father prevents weakening of the family's economic position in the next generation by subdivision of the land. The son who leaves agriculture thinks himself fortunate, whereas the one who stays behind feels injured.

A Bhathela background is undoubtedly still an important factor in the evaluation of members of the dominant caste in south Gujarat. A humble origin, however, may be partly or wholly redeemed by a number of more recent, chiefly individual, prestige criteria, such as income, level of education, and a nonagrarian occupation. Owing to the new scale of values the distinction between the two sections is blurred, certainly for outsiders, and the common features are stressed.[27] A common Anavil life style has gradually come into being, inspired, indeed, by the behavior pattern of the traditional notables, but also shaped by the adoption of various elements from the new urban and secular culture. It is abundantly clear that the Bhathelas of the garden villages are rapidly making up their disadvantage. The nine members of this caste who work outside the village all belong to the younger generation.

If the present evolution continues, will not all members of the dominant caste in the end have left the village community and taken up urban professions? That remains to be seen. Economic success has become a very important determinant for social status. The Anavils of Gandevigam, after all, attained prosperity through agriculture. As already stated, I am convinced that intensification of crop cultivation would yield even higher revenues. Yet even a rather inactive management provides the landlord with an income well above that earned by most of his fellow caste members in town. In other words, agriculture is so extremely profitable that it cannot simply be abandoned. By its revenues, moreover, it enables

27. Cf. Joshi, *op. cit.*, 108–10.

the landlords to realize their influence and prestige in both the "traditional" and the "modern" way — in buying, as it were, esteem for themselves by means of suitable marriage arrangements as well as in equaling their urban fellow caste members in education and material comfort.

A continuing transformation of the fields into mango orchards, and informal sharecropping arrangements with reliable Kolis and Dublas for the exploitation of the remaining land, will enable the *bagayat* Anavils to continue to draw income from their agrarian possessions. It is the perfect way out of the dilemma posed by their desire to concern themselves as little as possible with agriculture and at the same time to enjoy as high an income as possible. The most prosperous among them are considering whether to follow the example of one of their number and appoint a Koli as supervisor, which would accentuate their freedom from this humiliating labor.

Another factor of importance is that the garden villages have become much more agreeable to live in. Improved road connections have brought the towns in the vicinity within easy reach. More and more Anavils have their own means of transport; many have bicycles, and a few rich landlords of neighboring towns can afford an automobile. Housing in the villages has become much more comfortable and is, moreover, often cheaper than in town. Facilities, too have been improved; the installation of electricity and water has especially benefited the members of the dominant caste.

Those who combine a job in town with the management of their property have completely cast off the despised peasant existence. To this category in Gandevigam belong a number of Anavils who are primary or secondary schoolteachers and in their leisure hours keep a watchful eye on their agrarian interests. Their fellow caste members in the surrounding villages who daily go by bicycle or bus to town, where they have administrative jobs, are in the same position. To others they are shining examples. While profiting by the returns from agriculture they succeed in drastically limiting their dependence on it.

THE RELATIONSHIP BETWEEN ANAVILS AND DUBLAS. I could not have chosen a more desirable period for my stay in Gandevigam than the mango season. In the Anavil street, where I had been allocated an empty house, I passed many evenings in hospitable company. They were pleasant gatherings, usually on the veranda of the house of my neighbor, a large landowner. At a gesture from him, his servant, a Dubla boy, brought in a large basket of mangoes. These were the windfalls, the bruised or overripe fruits of the day. With some other neighbors, regular visitors like myself, we ate as much as we wanted. The servant stood in the background awaiting further orders. He would drag in another basket if the gentlemen had not had enough, and after we finished he would pick up the pits we had thrown down. Insofar as the servants are still allowed to take mangoes home at harvest time, the quantities are certainly smaller than they used to be. The servants, however, cannot afford to eat them, and sell them to a buyer from town who waits at the end of the day just outside the village.

The mango season largely coincides with the period in which most marriages are performed, and during the months I spent in Gandevigam I attended many weddings. In the most prosperous Anavil fashion these weddings are lavish occasions celebrated with a great deal of pomp and circumstance. In the marriage ceremony, which expresses all kinds of relations of reciprocity, superordination, and subordination, Dublas also take part. The contribution of servants and maids of the household consists, for instance in carrying the pot with earth from the bride's native village, and holding the jars from which water is sprinkled. These are symbolic of services that require physical effort, and the Anavils, having ensured the assistance of Dublas in their household, can demonstratively and to their greater glory abstain from them. Often many hundreds of guests partake of the marriage meal, but members of the caste of agricultural laborers are of course not included. The Dublas stood looking on from a corner until after the meal, when they could take the leftovers, chasing away the stray dogs of the village.

The foregoing paragraphs illustrate once again the contrast between the plentitude in which the landlords live and the want to which the landless laborers are reduced. Moreover, the two situations express the nature of the social relation between the two castes: the subordination of the Dublas to the Anavils. This applies in the first place to the farm servants, a numerous category in the garden village, but in a wider sense to all Dublas.

After the disappearance of the *hali* system the Dublas' subservience continued to exist, though no longer as a goal in itself.

That is to say, attachment served to maximize the income of the landlords, while the laborers were forced to work harder and they were used more effectively, at the same time receiving minimal remuneration. The Dublas' standard of living has fallen accordingly.

The Dublas never tire of discussing the egotism and greed of their employers. They are convinced, not incorrectly, that it is due to their effort that the members of the dominant caste can lead lives of luxury and idleness.

Conversely, the Anavils believe they are the ones who have reason to complain. In their eyes, the Dublas have changed from faithful, industrious, and obedient subordinates to recalcitrant good-for-nothings, dishonest and underhanded in their dealings with superiors. To the Anavils' displeasure, the Dublas stay away for days without notice, or work in secret for a Koli who pays them a little more. A recurrent scene in the village is that of an irate and scolding master on his way to fetch a servant for his daily job.

They frequently accuse the Dublas of theft, remarking wryly that the agricultural laborers work harder in their masters' fields during the night than they do by day. The neighboring towns contain some merchants who are known as receivers of stolen vegetables and fruit. By various means the landowners try to put an end to these practices. Some inspect the small bundles of the laborers when they leave the land in the evenings, many make nocturnal tours of their gardens. The large landowners recently hired a field watchman from another region during the harvest period of the chief crops, but this measure turned out to be rather ineffective. At one time during my stay there was a great commotion when it came to light that one of these Bayias was in league with a number of Dublas and an Anavil from the adjoining village.

The foregoing observations reflect clearly the fact that the judgment of the Anavils about the Dublas is in general negative. Agricultural laborers, they say, have the minds of animals — they take no thought for the future, are extremely uncivilized, have dirty habits, are stupid, and are unpredictable in their behavior.

On the other hand, several members of the dominant caste are aware that this judgment is severe and unfair. One of them remarked that, although the Dublas' alcoholism is always mentioned, the consumption of alcohol among the Kolis is certainly not less widespread. They are aware that they exploit the weak position of the agricultural laborers. These Anavils regret "the decline of the moral standard" during the last few decades. In their opinion, the result has been that the landlords think in terms of money alone, and that the agricultural laborers, for their part, do not wish to work. The decline began as early as the nineteen-forties. One Anavil told me that during the war period a loyal servant used to hand over his sugar-ration card of his own accord (the servant did not eat sugar anyway, he added). If the servant did not, however, and preferred to sell it in the black market, the master took it from him. That was wrong, just as, in his opinion, it would be wrong for the master not to give the servant any food, if it suited his convenience. The same applies to a landlord's unwillingness to give *khavati* and his refusal to help in case of illness.

Some Anavils frankly admit that they treat their laborers badly. "We look after our cattle better," one of them literally said. A landlord who conducts himself as a patron is praised by his fellow caste members. He should not belong too unmistakably to the type of working farmer, nor be too industrious and thrifty. A natural ascendancy over members of lower castes, combined with a gracious familiarity towards subordinates, is generally appreciated. Greed and brutishness are unworthy of an Anavil, as they are inconsistent with a landlord's pattern of behavior. Physical punishment belongs to the past, although a furious master may be excused if he kicks his servant now and then, and not all Dubla boys working in a household are spared an occasional beating. The landlord whose laborers do not run away wins admiration. A grasping, short-tempered master, on the other hand, makes himself ridiculous, for he has lost his ascendancy. "Good" Anavils therefore have no trouble at all in finding new servants. Their benevolent and accommodating behavior, however, is not due to compassion for the Dublas' predicament, but is practised as one of the virtues of a dominant caste to validate its supremacy. Lest I should be misunderstood, this is the recommended behavior. Reality is usually different. All Anavils wish to have cheap, hard-working labor, and most of them are indifferent towards the conditions under which the laborers live.

The landlords no longer even observe the unwritten rule of allotting the work as much as possible to Dublas from the village. At the peak of the season they hire labor from another subdistrict. These *biladia* are contracted as gangs for one season; they find their own food and build temporary huts in the fields and gardens. Another advantage in employing them is that because of their unfamiliarity with the local situation, they have no outlets for stolen fruits or vegetables. By making use of temporary labor in this way the Anavils show that they do not wish to accept any obligations over and above remuneration of the work performance. Employing casual labor on a daily basis relieves them of a great deal of "unnecessary trouble," but on the other hand weakens their grip on the Dublas.

The landless laborers have found, in their growing mutual alliance, a new means of defense against the Anavils. The Anavils themselves say that the Dublas, living together in the new quarter, have developed a "group instinct." The laborers left the gardens about ten years ago and bought pieces of land, just outside the village, on which to build their huts. Each of them paid the sum of 6 rupees and 3 annas, which went into the village exchequer. Because of the lapse of time, it was no longer possible for me to find out who had initiated the move. At any rate, it was not done against the will of the Anavils. Some of them feared they would lose authority over their subordinates, but most hoped that there would be fewer thefts of garden crops. The former expectation was justified, the latter was not. The Dublas have become a little less susceptible to the sanctions of the Anavils. The scattered location of the huts in the gardens had reinforced the Anavils' grip on their laborers. Now that the Dublas have settled together in the new quarter, their solidarity has increased. In the evenings after work they assemble in groups to talk, and inevitably the information they exchange fans the flame of existing feelings of dissatisfaction. The Anavils now not only have to reckon with the individual resistance of a subordinate, but in the last resort they may have to deal with the whole caste. At present, group control is primarily inward-directed. A man who has given in too easily to a superior's wishes is criticized by the other Dublas, sometimes even censured. It is, for instance, less easy now for the landlord to have sexual relations with the wife of an agricultural laborer. In the first place this is made difficult by the grouping of the Dublas' living quarters. Then, too, the farm servants themselves are under pressure from their own group to refuse to accept this behavior on the part of the master. Although I did hear of a few cases of heterosexual or homosexual relationship with subordinates, they appear to occur less frequently nowadays, and criticism of such relationships has become stronger.[28]

A quarrel between individual Anavils and Dublas may develop more easily than before into a larger conflict. When, a few years ago, some agricultural laborers were accused of attempted theft and were beaten by Anavils, a protest meeting was called in the Dubla quarter on the following day. A leader from the Praja Socialist Party in a neighboring town was invited to speak. It was not much more than a symbolic gesture of opposition to the unlimited authority of the landlords, and indeed was understood as such by the latter.

Another recent phenomenon is that the Dublas may go on strike. Some years ago the farm servants of Gandevigam stopped work for a few hours when they were denied the right to pick wood in the gardens and the landlords refused to raise their wages in compensation. This action failed. The Anavils fetched laborers from other villages and, to demonstrate their firmness, went to work themselves.

It may be concluded from the above-mentioned instances that protest has in some cases become an expression of group behavior. Does this mean that Anavil Brahmans as a group have become opposed to Dublas in general and that there is question of progressive intercaste antagonism? Joshi, in his study of a village of about the same composition in south Gujarat, does arrive at that conclusion. In my view, however, it should not immediately be concluded from the increasing antagonism between landlords and agricultural laborers that one or both parties present a solid front. I have described in great detail the continuing subservience of the Dublas in Gandevigam. Those who work outside the village are exceptions. Half the Dubla men are bound as farm servants. The dependence inherent in this relationship of service is utilized by them to obtain various benefits, amounting to credit in the broadest sense of the word: incidental aid, cattle to rear, land for sharecropping, a job for a son or brother in town, intercession with a third party. By granting favors of different kinds, or by holding out the prospect of favors, the members of the dominant caste maintain a system of individual alliances which at the same time keeps the Dublas divided among themselves. Day-laborers may be

28. Joshi arrives at a similar conclusion. (*Ibid.*, 107–08.)

absorbed into this network when they have taken up an advance and become indebted. The Dublas of the village compete with each other for the places and privileges which the Anavils have to distribute. Divergent, and partly conflicting, interests prevent the rise of a widely shared feeling of solidarity.

Unity no more exists among the members of the dominant caste than among the Dublas. The lack of concord among those in Gandevigam had been pointed out to me in advance by the Anavils of Chikhligam. This does not mean, as I thought it did, that the Anavil group in the garden village has broken up into opposing factions. Characteristic of them is rather a pronounced individualism. They oppose each other through indifference rather than because of willful contentiousness. Several Anavils no longer observe the unwritten rule of not taking on each others' agricultural laborers unless permission has been explicitly given. Some less prosperous masters have recently lost their farm servants to a larger landowner, who did not even compensate them for the Dublas' debts. "What business is it of mine?" they say. They are unmoved if their fellow caste members call them "thieves of other people's servants." No one, however, pushes matters to extremes. The garden owners look after their interests outside the village individually. Gandevigam is not an arena in which they fight out their differences of opinion. In general they pay little attention to each other's comings and goings, or rather, they reject undue efforts to influence their own acts and behavior.

The relations between Anavils and Dublas no longer have the patronage character of *halipratha.* On the other hand, the network of individual alliances controlled by the landlords has prevented the development of horizontal solidarity among the agricultural laborers. Although the latter are beginning to show some unity, no indissoluble solidarity or class consciousness has yet come into being.

The Dublas lack the power to improve their situation themselves. Resistance is limited to short-lived actions, and unity of purpose is confined to at most one village. As agricultural laborers they have hardly been touched by political activity. The parties of the left have thus far mainly directed their efforts towards the urban scene and, in the rural area of south Gujart, chiefly to the category of small farmers. At elections, the Dublas in many villages vote for the same political party as the landlords, usually the Congress

Party. Sometimes this reflects direct pressure exerted on subordinates, but often the mechanism works in a subtler way.

Key figures in the countryside are the so-called social workers, mostly members of the higher castes who, inspired by a mixture of social awareness and political ambition, stand up for the underprivileged and backward groups. Associating themselves with a social movement, usually of a Gandhian type, they are of interest because they control a bloc of votes in local political parties. Many of them are affiliated with the Congress Party, long the most important political entity in the area. Thus, although landlords and agricultural laborers often vote for the same party, they do so from dissimilar backgrounds and for different reasons. The "social workers" make themselves mediators in local disputes between the Anavils and Dublas. They urge the former to treat their subordinates better, and warn the latter against the dangers of a too militant attitude. When, despite their efforts, a conflict flares up, the mediators rarely take the side of those whose rights they claim to champion. Having learned by experience, the Dublas have little confidence in them. On the other hand, the "social workers" are among the few channels through which agricultural laborers can obtain a hearing in the higher realms of the state bureaucracy, while, in the Dublas' name, they maintain contacts with those government bodies that allocate favors. Four Dublas, who regard themselves as spokesmen for their caste in Gandevigam, have successfully used these channels and received subsidies through a "social worker" (an Anavil in a neighboring village) to build larger huts of better material, comparable to those in Chikhligam.

The government does not evince much interest in the condition of the agricutural laborers. Although a number of existing regulations bar arbitrary acts by landlords, this passive support rarely changes into active concern and aid for the Dublas. The government has not been able to guarantee the assistance and support which used to be institutionalized in the *hali* system. Employment is limited, and social security is lacking. The only progress that has been made in south Gujarat is in the housing of the Dublas on land of their own.

What little the government does, however, draws the reproaches of the Anavils, who would prefer the authorities to prohibit the seeking of work elsewhere or to forbid strikes. Either directly or through political channels the members of the dominant caste have access to the local government bureaucracy, and they know how to manage it. Officials visiting the village meet Kolis only rarely and Dublas in the area not at all. This makes it clear why "de-patronization" has placed the Dublas in an even more isolated position.

The rapid numerical growth of the Dublas, their powerlessness, the contraction of the agricultural economy, the limited employment outside it, certainly for them — all these factors point to progressive pauperization. The agricultural laborers react by passive resistance, the chief means of protest they possess. Their behavior is characterized only in part by docility and servility. They do not openly refuse to carry out the landlords' orders, but sabotage them. They do not want to work hard for a minimal daily wage, but will not say so openly. Their servility is exaggerated, their obedience is too demonstrative not to arouse suspicion. It is the impertinence of the powerless. Their behavior has all the features of what has been called a culture of repression.[29]

The conviction that things were better in the past does not imply that the majority of the Dublas yearn for the shelter and security of former times. To have these meant at the same time to admit and accept a far-reaching inferiority. They are all too aware that their low position in the social structure is closely connected with the dependence in which they have always found themselves. This subjugation, which detracts from the feeling of self-respect that has been growing among them, not only affects the farm servants but includes all the members of this tribal caste. Their continuing servitude fills the Dublas with a veiled but profound hatred against their lords and masters, the Anavil Brahmans.

29. Holmberg, 1967.

Chapter 9

Summary and Conclusions

9.1 Chikhligam and Gandevigam Compared

These two villages differ greatly as to the share of the population engaged in agriculture. In Chikhligam, for some decades, increasing numbers of both Anavil Brahmans and Dublas have sought and found work outside the village. For the Anavils, nonagrarian employment meant departure for urban centers both within and outside south Gujarat. Most Dublas, on the other hand, leave the village only temporarily and return every year, from necessity, before the beginning of the monsoon. Agriculture has remained the main source of income for the large majority of Gandevigam's population. Most members of the castes of Anavils and Dublas here are still landlords and agricultural laborers, respectively.

In the garden village, Dublas are mostly employed as farm servants. This category is also represented in Chikhligam, though in smaller number, and the day-laborers of both villages, who work all year in agriculture, form a minority among the agricultural laborers. The villages therefore differ mainly in the number of laborers engaged in agriculture throughout the year, not primarily in the nature of the prevailing agrarian labor relationship. However, owing to the seasonal migration of the majority of Chikhligam's Dublas, attached labor in that village is characterized by a less lasting relationship, and the Anavil Brahmans' control of their farm servants is less effective than in Gandevigam.

An important factor in explanation of the difference in agrarian employment is soil fertility. Because of the poor soil quality and indifferent use of irrigation, much of the village land in Chikhligam is not cultivated intensively, and agricultural labor is in fact confined to a few months of the year. This has greatly reinforced the urban orientation of the members of the dominant caste in the village. Partly because of the contraction of the crop system, a process that is far more advanced in Chikhligam than in Gandevigam, agrarian production is much lower than it could be if available labor and capital were fully utilized. The difference in extent of involvement in agriculture of the Anavils of Chikhligam and those of Gandevigam — also traceable to the time lag in the efforts of the Bhathelas of the garden village to attain the life style of the leading section of their caste — is reflected in the labor intensity of the agrarian cycle and in the number of Dublas who succeed in finding a livelihood outside the village. Alternative employment is, apparently, not an independent variable, and the vicinity of towns is not decisive for its realization. Inasmuch as, owing to its very profitable character, agriculture has remained the chief preoccupation of the Anavils of Gandevigam, they use their influence on the Dublas at times to obstruct any efforts of their laborers to find work outside the village. In Chikhligam, on the other hand, where the members of the dominant caste have changed their way of life, prefer urban professions, and are no longer actively engaged in agriculture, there was an early lack of employment for agricultural laborers, and, consequently, many of the Dublas were pushed out of the local economy.

The changed attitude of the Anavils resulted in depersonalization of their relationships with the agricultural laborers, who nevertheless did not become any less dependent. The economic distance between the two castes as a whole increased greatly. Differences in the prosperity of Anavils are more conspicuous than differences in the poverty of Dublas. In this latter caste, differentiation has hardly begun, whereas it is progressive among Dhodias and Kolis. The rapid demographic growth of the Dublas has aggravated their social insecurity and thus contributed to the continuance of the landlords' control of their actions. Although they prefer not to be attached, more Dublas are day-laborers from necessity than from conviction. Only those Anavils who are known for exceptional rudeness, unreliability, and greed have trouble finding new farm servants. Solidarity among the Dublas increased after they had moved to a village quarter of their own, but as yet there is no question of a common front against the local power elite; nor is such a development systematically encouraged by outside parties. Thus far, resistance has been incidental and limited to individual cases. In these respects the difference between the two villages is only one of degree. Although less weight is attached to agriculture in Chikhligam than in Gandevigam, the relations in both villages between

Anavil landlords and Dubla agricultural laborers are organized more or less on the same lines.

9.2 Parties to a Process of Depatronization

The disintegration of the *hali* system marked the disappearance of patronage. The possibility of selling an increasing proportion of the agrarian output on the market led to commercialization of the relationships between landlords and agricultural laborers. The allowance paid to Dublas, which in the subsistence economy had been based on various "obligations" towards the landlord, increasingly took on the character of a wage payment in exchange for a labor performance in the market economy, and it was paid in money. The amount was no longer related to the minimally defined needs of the agricultural laborers, but was based on the supply of labor and on the control exerted over them by the landlords.

The patron-client relationship has not disappeared in all respects, but the locus of patronage has shifted to a supralocal level and is framed in a more limited context, being chiefly economic or political in nature. The network of brokerage in the new situation, the parties in it, and its character are subjects which are outside the scope of this study. I have related "depatronization" to an enlargement of scale which put an end to the closed nature of the traditional social and economic order, but both the course of this process and its results were determined to a large extent by the reactions of the parties involved towards these outside influences.

THE LANDLORDS. There is no doubt that the landlords desired and promoted the disruption of the *hali* system. The change in the scale of preferences of the members of the dominant caste — high investments in education and in a consumption pattern adopted from the urban culture — was to the detriment of their subordinates. The changeover to market production made possible alternative ways of spending the crop profits and diminished the superiors' social responsibility towards their servants. Many Anavils seized the opportunity of cultivating fruit trees so as to free themselves from agricultural labor. Even without employing farm servants they can now lead a landlord's life as behooves a member of the dominant caste. The social gap between the Desai and Bhathela sections, which used to be extremely important, has narrowed as their economic positions have become comparable, with the consequence that their life styles are more alike. Urban orientation has enhanced the feeling of unity within the caste. Whereas, in 1921, 785 out of every 1,000 Anavil Brahmans were still landlords,[1] this ratio has drastically changed. Concerning the members of this caste, I. P. Desai remarks:

They are now convinced that not agriculture but salaried employment or occupations . . . are the means of livelihood. In their evaluation of occupation, "service" comes first and probably agriculture has no place. (Education is, for the Anavils, the suitable preparation for life in the city.) Now it is fixed from the day the boy goes to school that he has to be a salaried employee and they think of what he has to be and think of what they call the "line" for the boy. Education is becoming a normal necessity and so much so that they are sending their girls also to the secondary school. Some girls every day walk four miles to the school and back to the village. Even for a girl matriculation is becoming the minimum requirement.[2]

Their integration in the urban environment has given importance to other standards of values than the traditional ones, especially as to education, income, and a profession outside agriculture. The village community is now only in part the framework in which the position of the Anavils is determined. The dependence of members of other castes is no longer an aim in itself, but has become a means to enhance the masters' own prosperity and to finance preparation of their children for nonagrarian professions. Ostentatious leisure is now possible without a landlord's life, and the desire for conspicuous consumption is satisfied by the display of riches. Even where the landlords are still actively concerned in agriculture, the grip of the members of the dominant caste on the Dublas has become more depersonalized, being dictated by income maximization. The agricultural laborers' right to work is no longer recognized, their social security no longer guaranteed. In the endeavor of the members of the dominant castes to attain more esteem and influence within and outside the village, the Dublas have been changed from subjects into objects.

THE AGRICULTURAL LABORERS. Descriptions of the disappearance of the *hali* system often create the impression that it was the agricultural laborers who initiated it — that when alternative em-

1. *Census of India*, 1921, vol. XVII, Baroda, pt. 1, 363.
2. I. P. Desai, 1966, 81.

ployment arose, increasing numbers of them severed the relationship of servitude. My data confirm this assumption only to a limited extent. The need of the local landlords determines whether any Dublas succeed in finding work outside the village, and if so, what proportion of them. Once a migration tradition has arisen, as in Chikhligam, this changes. Contact with alternative sources of income in the cities then takes place through the Dublas' own mediators, the *mukadams*, and is no longer controlled by the Anavils. Even then, escape is often only temporary, because of the unskilled, low-paid, and unstable character of urban employment that is available to Dublas. The termination of the traditional service relationship may also be connected with the Dublas in another way. It might be argued that the rapid growth of their number makes it impossible for the landlords to provide for them all. This fact undoubtedly reinforced the effect of "depatronization," but it did not cause it. The agricultural economy of south Gujarat could have been much more labor-intensive if the crop system had been different. The abundant supply of labor, in other words, is not traceable simply to the rapid demographic growth of the population of agricultural laborers, but is also connected with the preference of the dominant landowners for pursuing a kind of agriculture that is not highly intensive.

The elements of bondage that still exist are especially apparent in attached labor. Weber, who himself studied a similar shift in the relationship between landlords and agricultural laborers in eastern Germany in the late nineteenth century, found that farm servants in their turn began to resist their continuing servitude.[3] The same applies to the Dublas in south Gujarat. A fundamental difference exists, however, in that the rise of alternative employment in India has been much slower, whereas the population is increasing much more rapidly than it did in Germany in the late nineteenth century.

Although indeed aspirations have been aroused among the Dublas that are incompatible with the bondage inherent in attachment, with dependence, and with subordination in general, it seems that the members of that caste were above all subjected to the process of depatronization. In more than one respect they were the losing party.

3. Weber, 1892, 797–98.

EXTERNAL INFLUENCES. Factors other than those mentioned above played a role as well. In many government reports the *hali* system was criticized. Steps were taken to abolish it, but they started from the assumption of debt slavery and, if only for that reason, were not very effective. These regulations, moreover, were too half-hearted to allow of serious implementation, even if the local bureaucracy had cooperated. The governmental sanctions were chiefly effective in a passive sense, the semblance of legitimacy having been removed from the arbitrary actions of landlords towards runaway or otherwise "disobedient" farm servants. Administrative intervention from outside, a very important component of the enlargement of scale, affected the traditional authority of the local power elite. Yet the fact that the government administration was made more autonomous — that rules and procedures were standardized and the staff of officials gradually became professionalized — could only offset in part the influence of the dominant caste. In the first place, members of that caste had been incorporated into the local administration as village headmen and police *patels* from the early nineteenth century onwards. Moreover, in their entry into the higher echelons — which was subject to education and connections — the traditional elite, in the course of time, had built up a lead which greatly enhanced their influence.

These, in brief, were the main reasons why the regional bureaucracy always accommodated itself to the interests of the dominant caste, and also why the local administration made few if any efforts to improve the condition of the agricultural laborers. The use of brute force by landlords is, of course, illegal, but it is tacitly sanctioned by the authorities.[4] Only such intermediate castes as the Kolis in Gandevigam have succeeded to some extent in shaking off the yoke of the dominant landowners, but in a struggle for power the former can rarely hold their own against the latter.

Political movements have had no more important influence than has the government. Since the nineteen-twenties the leaders of *ashrams* in the neighboring Surat district have striven for the abolishment of the *hali* system, to which end they called upon Gandhi, Sardar Patel, and others for help. An agreement to this effect between the landlords' representatives and agricultural laborers was ratified in 1939 by Gandhi, though not without demur, for he thought the conditions of the agreement favored the land-

4. Béteille, 1965, 182–84.

lords and were disadvantageous to the laborers. The main stipulations were that the Dublas would thenceforward receive their wages in money, and that their debts would be canceled. The agreement was confirmed some years after the Second World War. This well-meant intervention, however, resulted in a deterioration of labor conditions, for since then the amount of wage, rather low even at the time, has not increased correspondingly with the price rises of the articles that the Dublas used to receive as wages in kind. Reduced consumption was the immediate result. The leaders of the emancipation movement in the region have not unequivocally declared themselves against the dominant landowners; they try to bring about a change of mentality among this local elite by persuasion and collaboration. Moreover, the *ashrams*, whose leaders belong to high castes, are greatly influenced by a Hindu doctrine that rejects some tribal culture elements in Dubla life. In moralizing and even reprimanding terms, the members of tribal castes are lectured by social workers about all kinds of "deviations" in their social life that are looked upon as inferior, and are censured especially for their lack of hygiene and their consumption of alcohol. Rather unrealistically, the Dublas are advised to avoid contracting a debt, as they lose their freedom in this way. Under these circumstances it is not surprising that the ideas propagated by the *ashrams* scarcely take root, and that social workers attached to these institutions enjoy the confidence of the tribal castes only to a limited extent. They have not been able to bring about any change in the existing relationships of power and property by their efforts. It appears from the report written by the leader of this movement in the northern *talukas* that the landlords took little notice of the pressure exerted on them to treat their agricultural laborers more humanely.[5] From its beginning, the emancipation movement, which spread gradually to other parts of south Gujarat through *ashrams* and social workers, confined itself to persuasion and reconciliation and did not essentially affect the social system in which landlords and agricultural laborers function.

9.3 The Present Situation

What relations between landlords and agricultural laborers have replaced the *hali* system? — this was the second question that I posed at the beginning of my field work. The labor system in a market

5. Dave, 1946.

economy is primarily determined by labor supply and demand. Whereas in some regions the demand has remained about equal, and in other parts of south Gujarat has considerably shrunken, the caste from which the laborers were traditionally recruited has grown rapidly. Opportunities for employment outside agriculture have not kept pace with these developments. The kind of work available to Dublas in the urban centers reduces them to a life of pauperism, and their condition is not much better than that of the rural Dublas, whose labor conditions are imposed by the landlords. The living standard of the agricultural laborers, low to begin with, deteriorated further when allowances in kind were replaced by money wages.[6] Most of them are continually indebted, and this is the main reason why, in budget calculations of households of agricultural laborers, expenditure always turns out to exceed income.[7] The debt binds the Dublas and provides the landlords with a means of pressure. In the long term it is cheaper and safer for the landlords to give a limited loan to a farm servant than to hire day-laborers. Moreover, they select the most industrious and obedient laborers for farm service.

The mutual rights and obligations have always been one-sidedly weighted. I pointed out in an earlier chapter the disguised exploitation that could be perceived in the *hali* system. On the other hand, patronage-like elements still occur in the present relations between landlords and agricultural laborers. I know of a number of Anavils who still feel obliged to aid and protect their subordinates, and these servants, in their turn, sometimes speak of their masters in terms of affection and devotion. As for the landlords' endeavor to acquire political support within the village for the purpose of obtaining advantages outside, it might be described as patronage for new ends.[8]

To the conclusion that exploitation was inherent in the *hali* system, and that patronage may occur in a situation of exploitation, should be added the comment that the difference between past and present is more than one of degree.

It would, therefore, be incorrect to interpret as a form of patron-

6. Concerning the Dubla agricultural laborers in his field village, I. P. Desai writes, "They live from day-to-day and even by any standard starvation among them is general, even in 1963, in spite of the five-year plans." I. P. Desai, 1966, 140.

7. By adding to their income the amount by which the debt of the agricultural laborers increases every year, this misunderstanding would be solved.

8. Cf. Béteille, 1965, 182–84.

age the network of obligations with which the Anavils have surrounded themselves. Although the landlords wish to preserve their traditional rights, they are not prepared to accept the attendant responsibility for the well-being of the Dublas who depend on them. On the contrary, they try to limit their obligations to the utmost, that is to say, to specify them and reduce them to a minimum. In other words, the Anavils do not behave as patrons, but still expect a maximal benefit of varied counterservices. Their demands increase as the dependence of their subordinates grows. All agricultural laborers among the Dublas depend on them, but dependence is particularly inescapable for farm servants. They are bound to about the same extent as the *halis* were, but they do not receive the aid and protection that their predecessors enjoyed. The farm servants in their turn therefore try to confine themselves as much as possible to a fixed amount of work, to render their obligation specific instead of diffuse. Whenever possible, they sabotage any claims by the master that exceed that amount. Both parties have exorbitant expectations towards each other, but they try to get away with fulfilling partial roles as employer and employee, respectively. In the end the Anavils, by virtue of their dominance, succeed better at this than do the Dublas. The risk of subsistence has been shifted onto the agricultural laborers. The distinction made by DeVos between expressive exploitation on the one hand (in which the mere fact of recognition of dependence by others creates satisfaction and is a purpose in itself) and instrumental exploitation on the other hand precisely illustrates the change which has taken place in the relationship between the landlords and agricultural laborers.[9] The widening social gap is even harder to accept for the Dublas than is the growing economic disparity. The Anavil who used to work alongside his subordinate now keeps aloof from agricultural labor. By freeing himself as much as possible from his undignified work the landlord has at the same time made the landless laborers more conscious of their humble position.

The Dublas are increasingly aware of the fact that their subjection is one of the main causes of their low social status. Economic disadvantages aside, weakness and dependence are looked upon as dishonarable and deplorable.[10] Their resistance takes various forms.

9. DeVos, 1966, 68–72.

10. Cf. Barth, 1960, 140, and Harper, 1968, 37. Also Weber, 1892, 797, where he says, "Dissociation from the patriarchal community of home and economy at any price, even at that of passing into the homeless proletariat, is the most marked

They no longer submit to beatings, nor do they tolerate landlords maintaining sexual relations with Dubla women. In the tense situation that now exists, incidents of this nature occasionally lead to sudden and violent explosions.

To sum up, however, the agricultural laborers cannot escape the control of the dominant landowners. Attachment as farm servants symbolizes the dependent condition in which more or less the great majority of Dublas lives. They still depend on the landlords for their livelihood. Those who have escaped and have found work outside the village and agriculture are highly esteemed by their fellow caste members but are looked upon with great irritation by the local power elite. When the members of the dominant caste oriented themselves towards the urban environment while maintaining their interests in the village community, their indifference to the fate of the Dublas became even more evident.[11] Dependence in an economic sense is used by the landlords to further political ends as well. This explains why the Anavils, though not very numerous, have largely been able to maintain their supremacy in the rural area of south Gujarat. They mobilize the support of the members of the lower castes to legitimize their dominance, and thus to perpetuate it in the new political structure.

The process of "depatronization" has left the agricultural laborers in a condition of isolation. The landlords are averse to being reminded of their former obligation to look after the interests of their subordinates, and sometimes flatly refuse to intercede for them with a third party. Nor are there any indications that the Dublas are trying to join forces with other weak categories, for instance the small farmers. The basis for such a collaboration is lacking.

The government has not taken over the functions of patronage, especially those of protection and social security. Industrial employment is expanding at a slow pace, and so far has not been sufficient to counteract the increasing pressure on the agrarian sources of existence. A proposal on the part of the central government to introduce a minimum wage for agricultural laborers was delayed for a long time; in the end, most states did not adopt it, and the

tendency in precisely the ablest elements among the laborers. . . . It is the tremendous and purely psychological magic of 'freedom' which finds expression in it. It is largely a question of a grandiose illusion, but after all, man, and thus the agricultural laborer, does not live 'by bread alone.'"

11. See also Béteille, 1965, 201–02.

remaining ones obstruct the implementation of provisions to this end. Nor do the states enforce compulsory education, which was introduced as an important medium for the improvement of the economic and social position of backward groups. Not only the older Dublas but a large segment of the younger generation are illiterate, and only thanks to the activities of an emancipation movement is this situation beginning to change locally. On a modest scale, the government has enabled the members of this tribal caste in south Gujarat to settle on land of their own in the villages, and also occasionally has subsidized the construction of huts on the land. The amount of money allocated to this project is low, however, and moreover it was not completely spent in the past few years. The government is perfectly cognizant of the miserable existence of the agricultural laborers. The commission appointed in 1947 to enquire into the extent of bondage in which the Dublas lived was instructed to draft proposals "enabling them to live a life consistent with human dignity and self-respect." The official reports concerning the backward regions and groups which have appeared since then, however, state that little has been done for the Dublas and that their circumstances have not changed.[12]

The Dublas have not succeeded in turning their numerical strength into political power. Nor has mobilization of their dissatisfaction by political parties yet been attempted. Especially in the Surat district, communist leaders have tried in recent years to win agricultural laborers over to their party, but in the state of Gujarat this party is too weak to have any influence. The votes of the Dublas in the villages are controlled by the landlords or by social workers. The emancipation movement that is operating in south Gujarat has founded a society to promote the welfare of Dublas (*Halpati seva sangh*), with a leadership consisting mainly of social workers and landlords who are affiliated to the Congress Party. At a conference that I attended, for instance, the Dublas were exhorted by the guest of honor, Morarji Desai, to break their bonds of dependence and not to submit to the bad treatment that falls to their lot, but at the same time they were urged to be patient. To control the increasing protest and rebelliousness seems to be the main function of this organization, but stabilization of the existing relationships amounts, for the Dublas, to stagnation, if not to deterioration of their condition.

12. See, for instance, *Report of the Scheduled Areas and Scheduled Tribes Commission, 1960–61*, New Delhi, 1961, 384.

The landless laborers have no access to the centers of power, and the isolation in which they find themselves is part of their economic and social pauperism. If we take their rapidly growing numbers into account as well, a prognosis cannot be other than pessimistic. It seems unlikely that the Dublas will soon know better times. Legislation which will guarantee a higher wage for the agricultural laborers is inadequate, as are other measures aimed at bettering their lot. Government intervention to this effect may aid in improving their material circumstances, but it must include economic and social reforms that will reinforce the position of the Dublas in the labor market. Industrialization, the change which in the late nineteenth century in Germany interrupted the process of proletarization of the agricultural laborers in the eastern areas, as described by Weber, progresses too slowly in India to provide a solution. The ratios between urban and rural population and between agrarian and industrial employment in south Gujarat have not appreciably changed in recent decades. This means that Dublas increasingly compete for the available labor places in agriculture — to the greater comfort and gain of the landlords.

Wider employment in agriculture appears to be a more realistic possibility. The much-discussed "green revolution," a program of measures in the field of agricultural economy and technology, which has recently led to a spectacular rise of agrarian production in several regions, also could enhance, and sometimes has enhanced, the labor-intensity per land unit. This does not, however, necessarily imply a rise in the real income of the agricultural laborer. There are, on the contrary, many indications that, because of the green revolution, the economic and social gap between landowners and landless laborers in the rural areas has widened further.[13] Improvement of the condition of the Dublas can be expected only if there is a fundamental change in the relationships of power and property, but the possibility of this is remote. The administrative decentraliza-

13. Cf. also the statement of Vyas in a report on the conditions of agricultural laborers in four villages of west India: "In most of the regions there are no organisations of agricultural labourers which can represent them adequately before the employers and secure for them higher wages and better conditions of work. The attitude of the State in this field is reflected more in pious hopes than in any active steps. The way the provisions of the Minimum Wages Act for Agricultural Labourers are flouted in almost all areas is a significant indicator. Because of all these factors it is quite possible that though the demand for agricultural labour may show an increase in relation to supply, the wage rates may remain depressed, as has been happening for the last decade or so. In such circumstances, even if there are no large-scale disturbances, a situation of accentuated tension may arise in many areas." Vyas, 1964, 23–24.

tion that has been introduced, as well as the consequent tendency of the local government bureaucracy to become politicized, have had, on the whole, advantageous consequences for the dominant landowners.

Among the Dublas, there is increasingly a collective awareness of their condition of subjection, a feeling of being wronged, and the beginning of an attempt at collective opposition against their exploiters. But it must be added immediately that the conditions that are prerequisite to implementing their protest organizationally in an effective and lasting way are lacking. Their rebelliousness is of short duration and is often directed against individual landlords. Like their companions in misery elsewhere, the Dubla agricultural laborers have become sunken in apathy and suspicion. Their lives offer them no prospects of improvement and they show themselves indifferent to what the future may bring. It can never be better than it is today — and that is not enough.

Chapter 10

The Findings Reexamined

To what extent are the findings of my investigation generally valid? Chikhligam and Gandevigam were not chosen arbitrarily; they are typical of a great number of villages in south Gujarat having a similar social and economic composition. In generalizing my conclusions, however, I must make one reservation.

In most of the villages of the northern plain of south Gujarat, Kanbi Patels as well as Anavil Brahmans are the dominant landowners employing Dublas as agricultural laborers. An insight into the nature of agrarian labor relationships in this area, the Surat district in its present form, would therefore yield an answer to the question whether my findings are indicative of castes of landlords other than the Anavil Brahmans, that is, whether it would confirm my assumption that it is economic position, not caste membership, that determines the way in which agricultural labor is employed.

Moreover, such a comparison would make it possible to devote greater attention to the political background of changes in the relationships between landlords and agricultural laborers. In Gandevigam and Chikhligam, each with its own administrative past, politics were not a factor in the disappearance of the *hali* system. Informants assured me, however, that politics did play a role in the northern *talukas*, where there was an active emancipation movement along Gandhian lines. In this region, I was told, it was through political action that *halipratha*, denounced by Gandhi as immoral, fell into discredit.

To check my earlier conclusions, therefore, I spent a brief period in one of the villages in the Surat district.

THE AREA AND THE VILLAGE. The villages in those *talukas* that have formed the Surat district since 1964 are situated in a plain of black and fertile soil, the basin of the Tapi River. A railroad and a bus line serve to maintain contacts within the area and with the city of Surat, the center for the whole region. The village that I selected in the eastern part of the plain, and which I will call Valodgam after the subdistrict to which it belongs, has about 600 inhabitants, most of whom are engaged in agriculture. This applies particularly to the members of the two largest castes, Kanbi Patels and Dublas, respectively landlords and agricultural laborers. In addition, several artisan and serving castes are represented by one or more households, whose members carry out the traditional crafts and skills (as blacksmiths, carpenters, shoemakers, hairdressers, etc.) or are employed in new occupations in nearby towns.

In Valodgam, as in many other villages of this region, Kanbi Patels, being the most important landlords, are in a dominant position. Wherever members of this caste settled — tradition has it that they originated in central Gujarat — they succeeded in monopolizing most of the arable land. Dispossession of the tribal population living in the region had already begun before the inception of colonial rule. With few exceptions, the Dublas who live in Valodgam are completely landless. The adjoining village is inhabited exclusively by members of the tribal castes, especially Gamits, but also Dublas, most of whom have held their own as small landowners. Yet in the past the inhabitants of this tribal village also lost a large part of their land, which was taken by Kanbis of Valodgam. Moreover, many of the tribal cultivators came to depend as tenants on Banias, who for a long time have inhabited the small market towns of this region. Among the autochthonous tribal groups there is an awareness that they have been partly or wholy dispossessed by these incoming elements. During recent years their protest has been expressed in a local movement of unrest, socioeconomic in nature and with a religious undertone. In this study, however, the data I gathered concerning this *sati-pati* movement are left out of consideration.

The plain in which Valodgam is situated is an important agricultural area. Towards the end of the last century, cotton was introduced as a market crop. Besides food grains, the growing of groundnuts (peanuts) in particular has gained importance. All of these are cultivated in Valodgam. Only the dry and hot summer is a period of rest in the agricultural cycle. On the low-lying *kyari* soil, rice is grown during the monsoon, and pulses (especially *val*) afterwards as a second crop. On the higher fields, the farmers grow millet, cotton, and groundnuts.

Employment outside agriculture in the region itself, which has no large towns, is not very common. In some small towns along the railway, a few workshops have recently been established, though as yet their output is not large and they employ only a few hands. Larger industries are to be found exclusively around Surat, and owing to improved transportation the pull of this city on the inhabitants of the inland territory is increasing.

THE LANDLORDS. The time at my disposal was too short to permit collecting information on the situation of all the Kanbi Patels, and I confined my researches to ten households of this caste, approximately one-third of the total number. Of these households, two possessed less than ten acres of land, four had between ten and thirty-five acres, and the remaining four had more. The largest landowner, who was also the village headman, actually owned over 100 acres. Although in this sample the large landlords are overrepresented, it is a fact that the average land holdings of the other Kanbi Patels are also rather extensive. Most of them have much larger herds of cattle than do the Anavil Brahmans. The village headman even boasts a horse, the only one for miles around, on which he hurried to the Dubla quarter as soon as I had been seen there.

Together the ten Kanbi households employ fifty-five farm servants, an important indication of the intensiveness of agriculture and the large size of their properties. Of this number, nineteen are in the employ of the headman alone. The permanent servants comprise cowherds, house servants who at the same time are agricultural laborers (*chakars*), and field laborers. Besides these, most households employ one or more maids. Agricultural laborers are chiefly Dublas from the village itself, but some also come from the adjoining village, inhabited by tribal castes.

The large and spacious brick houses along the village street have tiled floors as well as beautifully decorated wooden front doors. The rich furnishings bear witness to the wealth of the occupants, the Kanbi Patels. The members of that caste in Valodgam all belong to the Leva section, a dominant group in the villages of this region. In the past they were peasants who tilled the land together with their laborers. During and after the Second World War they became prosperous, and now many of them no longer do any physical work on the land. The fact that their prosperity is of recent date, however, does not imply that the previous generation were very poor. As early as the turn of the century, the Collector of Surat wrote about the Kanbis:

> The present men, as they live less laborious lives, so they have more expensive tastes than their forbears and to gratify them will resort to the *savkar* [moneylender] if there is no money in the house. Formerly the ordinary cultivators to a man wore country cloth; now they must have it of finer texture from Manchester. Cheap local rice, *dal* and *gul* [*gur*] were enough for the daily food; now vegetables, imported rice and refined sugar are in demand. A more luxurious generation seeks after *pan-supari* [betelnut], cheroots, hired servants, sweetmeats and American watches.[1]

A number of old district reports also mention the *hali* system. Servitude is certainly no recent phenomenon in the area, and the Kanbis of Valodgam were among those who at an early date employed *halis* and completely controlled them.

Now they feel hampered in the exercise of their power by the activities of social workers. At a short distance from the village there is an *ashram*, founded some decades ago, which since then has been a center from which attempts are made to improve the condition of Dublas and members of similar castes. This institution has been closely associated since it was founded with the Congress movement, the strongest political current in south Gujarat. Although the Kanbis of Valodgam were zealous supporters of Congress in the struggle for independence, they lost their confidence in the party because of the activities of the *ashram* workers among the tribal groups. They transferred their sympathy and political support to the Swatantra Party, which is popular among the large landowners in other parts of south Gujarat as well. The Kanbis are convinced that the Dublas are being set against them and are incited to leave their masters. They fear an increasing political awareness among their subordinates, but nevertheless still feel that they are lords and masters in the village, able to rebuff any "subversive" activities on the part of outsiders.

Although, like the Anavil Brahmans in the more southern and western *talukas* of south Gujarat, the Kanbi Patels are landlords, they have remained much more involved in agriculture, and are as yet only slightly urban-oriented. The younger generation attends

1. Broomfield and Maxwell, Bombay, 1929, 65.

schools and universities, and an increasing number of Kanbis are engaged in nonagrarian occupations in Surat or have left the country for East Africa or England, but this is especially to prevent landownership from being excessively fragmented among succeeding generations. In general the Kanbis have remained agriculturists much more than have the Anavils. They manage their property in a businesslike way and are solidly integrated in the village community.

THE AGRICULTURAL LABORERS. As mentioned earlier, the landlords employ large numbers of laborers, more than are available in Valodgam. My data, however, concern only the landless Dublas from the village itself. Of the men in the productive age group, thirty-six work as field laborers (here still called *halis*) and eleven as *chakars*, and six boys are *govalia*. Apart from these agricultural laborers in permanent service, twenty Dublas are day-laborers, and seven own a piece of land themselves or are illegal sharecroppers for Banias from a nearby town, in some cases both. Inasmuch as the tribal peasants from the adjoining village supplement their income from land by periodically working for Kanbis as well, these figures do not give an exact picture of the total employment pattern. Except for some scarcity at the peak of the season there is an ample supply of labor.

Govalio, chakar, and *hali* used to be successive phases in a Dubla's service relationship with a master. He began as a cowherd and after some time became a *chakar*. A few years after his marriage he usually entered the category of *halis*. Such careers still exist, but are no longer the rule. Moreover, a servant changes landlords much more often than in the past. In recent years the Kanbis have had some difficulty in finding new stableboys or house servants. Owing to the extreme dependence of such servants, who carry out part of their tasks in the master's house and have to be at his disposal from morning until night, most of the Dublas now become field servants without having passed through the preceding stages.

Stableboys (*govalia*) are as a rule young Dublas about fourteen years old, who tend the cattle and if necessary help out on the land. They live in the house of the Kanbi and receive, besides their meals and clothing, an annual wage of about 75 rupees. After some years they become *chakar*. They work on the land but also perform various jobs in and around the house. Their daily task begins at 4 o'clock in the morning and ends at about 9 in the evening. The unmarried house servants receive the same amount of food and clothing as the stableboys, but a higher annual wage, ranging from 100 to 125 rupees. Often this wage is not actually paid, but is partly or wholly entered in the books as owing to the servant until he marries. Once married, he begins to urge the master to relieve him of service in the house and restrict his task to agricultural labor, for which he will be paid a daily wage instead of an annual one — changes that are decided improvements for the Dubla. The Kanbi may try to delay this transfer to the category of field servant as long as possible, but if he does not wish to lose his servant he will finally have to give in.

Most of the attached laborers are farm servants who mainly work on the land. A number of them have been bound to the landlord since their youth; for others the attachment begins at the moment that they receive an advance from a Kanbi in order to marry. It has become very difficult for a Dubla to obtain a large initial loan, for instance some hundreds of rupees. This is not because the landlords fear that they will not be repaid — in fact, they know that repayment is practically impossible for the farm servants and are under no illusion in this respect. On the contrary, they are well aware that their claim will increase in the course of time. What worries them is the possibility that the local government will once again cancel the debt, so that the Dublas in permanent service will be released from their obligations towards them. The Kanbis must then make new advances of money so as to be able to exert pressure and ensure the labor of their subordinates, just as in the old days when most of the farm servants, after their debt had been remitted, soon put themselves in their masters' debt again.

The field servant's working hours vary according to the season, but they are always less than those of the house servant. As elsewhere in south Gujarat, the servants formerly received their wages in kind: daily meals and a grain allowance, and periodically clothing and other perquisites. After the agreement achieved some decades ago by the Gandhian workers, the customary distribution of small quantities of the produce at harvest time was ended. Buttermilk (*chas*), for instance, one of the few protein-containing components of their diet, used to be given free. Now the laborers have to buy it from their landlords, and most of them can rarely afford to do so. Meals are provided only on very busy days, when the farm servant works from 6 o'clock in the morning until 7 at night. At the time of my stay the daily wage was 1 rupee. On the request of their laborers, some Kanbis resumed distributing, on some days of the week, the traditional amount of grain instead of money.

The day-laborers in Valodgam form a residual category among the landless Dublas. They are those who have not (yet) managed to find a master, and their existence is precarious because they are not assured of work every day. The wage they receive is not higher than that of farm servants, and their chances of getting credit are even slighter.

In the region itself, alternative employment scarcely exists, and it is difficult for the Dublas to obtain such jobs farther away. The employment of a Dubla man from the village in a small town in Saurasthra set in motion the seasonal migration of some of his fellow caste members to the same town. The members of a few households migrate each year to the brickyards, and a small number of Dublas have succeeded in finding permanent employment in Surat and have left the village permanently. All in all, agricultural labor has remained the chief source of income for the members of this caste. That is to say, in the agricultural pattern of Valodgam the demand for Dubla labor is such that relatively few of them dare to take the risk of seeking work elsewhere.

A striking contrast between the Dublas of Chikhligam and Gandevigam and those of Valodgam is that many more of the younger members of this caste in Valodgam receive education. This is certainly due to the effort of the social workers here. Boarding-schools are attached to the *ashrams*, where boys of tribal castes, and increasingly girls, are schooled. Nevertheless, this improvement, however important socially, has not led to an improvement of the economic position of the Dubla agricultural laborers. The attempt to start a house industry by distributing spinning-wheels is completely in accordance with the Gandhian philosophy on which the *ashrams* are founded, but has not sufficiently taken hold among the tribal population to be of any real importance.

However, their activities in the educational field did have the result that the tribal groups have become aware of their dependence much earlier in Valodgam than elsewhere in south Gujarat. They gave a content of their own to their protest in the *sati-pati* movement mentioned above. The leaders of the movement were Gamits and Dublas who had been educated in the local *ashrams* but turned away from them in disappointment.

THE RELATIONSHIP BETWEEN LANDLORDS AND AGRICULTURAL LABORERS. The results of an investigation carried out by the agricultural economist J. M. Patel in 1961–62 into the condition of the agricultural laborers in a village near Valodgam largely agree with my own findings.[2] It appears from this report, which only recently came into my hands, that agricultural laborers, practically all Dublas, devote 86 percent of all their expenditure (an average of 593.15 rupees per household) to food, and not more than 3.5 percent to "social and miscellaneous expenses," whereas the Kanbi landlords allocate 46 percent of their expenditures (an average of 4,210.63 rupees) to food, and have close to 33 percent left for "social and miscellaneous expenses."[3]

The fundamental pattern of the relationship between Kanbi Patels and Dublas in Valodgam is identical with the situation in Chikhligam and Gandevigam which was described in preceding chapters. The agricultural laborers are conscious of servitude as a form of attachment that detracts from their own dignity. The rude treatment that is often their lot emphasizes their subservient condition. This applies especially to the *govalio* and *chakar*, who work in the master's house and are continually in his vicinity. For the Dubla, that security of regular work and the possibility of obtaining small loans outweigh the disadvantages of dependence. The Kanbis manipulate their subordinates' debt as means to exert pressure but, taught by experience, they try to keep the amount as low as possible. They grant a larger loan only when an agricultural laborer has worked on probation for a year, and then only if another Dubla, one who has proved to be reliable, will stand security for him.

In Valodgam, too, the changeover to the money wage was part of a more comprehensive process of commercialization. The Kanbi Patels conduct themselves like managers, running their land prop-

2. J. M. Patel, 1964, 92–119. A difference is that Patel defines the field servants among the Dublas as "casual labourers"; this is misleading, as his own description of this category shows (*ibid.*, 107). He only refers to the Dublas in the category of *chakars* as "attached labourers."

3. Patel does not mention how he collected these data, nor does he relate them to the incomes of the households. However, even if we do not attach too absolute a value to the amounts below and to their mutual ratios, his figures are useful indications of the differences between living standards of landlords and those of agricultural laborers.

erty in an efficient manner. They complain of the high labor costs and take on no more agricultural laborers than are strictly necessary, but if it suits the landlords better, they too take on laborers from elsewhere. They no longer feel obliged to give the Dublas of their own village an opportunity to work and provide for their existence. Without a doubt the rapid rise in the number of members of the Dubla caste has contributed to their indifference.

Relations became strained with the spread of commercialization. The personal tie of affection between master and servant is tenuous now because the relationship often begins at a later age, and severance occurs more frequently than it formerly did. Dublas

TABLE 10.1

PERCENT EXPENDITURE ON VARIOUS ITEMS BY ANNUAL HOUSEHOLD EXPENDITURE OF AGRICULTURAL LABORERS' HOUSEHOLDS, LANDOWNERS' HOUSEHOLDS, AND AVERAGE VILLAGE HOUSEHOLD

Item	*Percent of total expenditure of agricultural laborers' households*	*Percent of total expenditure of households of cultivating group*	*Percent of total expenditure of all households in village*
Food	85.94	45.59	56.85
Clothing	6.84	7.29	7.56
Education	0.16	1.89	1.44
Fuel and light	1.92	4.90	4.20
House rent and repairs	.33	1.25	1.43
Health and recreation	1.14	3.62	3.30
Traveling	.14	1.84	1.50
Social and miscellaneous expenses	3.53	32.62	23.72
Total	100.00	100.00 (read: 99.00)	100.00
	Agricultural laborers	*Landowners*	*All*
Total annual expenditure per household (rupees)	593.15	4,210.63	1,624.28

complain of the greed, exactingness, and hardness of their masters, and in their turn appear to be marked by the same objectionable characteristics as those of their fellow caste members elsewhere. The Kanbis call their subordinates lazy ("they come to work only five days of the week"), dishonest, secretive, and impertinent. Their disobedience and rebelliousness taxes the patience of the masters, but ultimately it is the Dublas who come off worse.

As in Chikhligam and Gandevigam, the nostalgic way in which both parties in Valodgam speak of their former association does not ring very true and seems primarily to reflect the friction characterizing their present relationship. Nevertheless, the conflict of interests between them has undeniably increased with time. The Dublas are convinced that in the past they were better off, and this view is confirmed by the local social workers.

The pattern of relationships between landlords and agricultural laborers has changed in Valodgam in the same way as that described above for the two other villages. It is true that the emancipation movement in the region has given this process a particular content, but it has not much affected its direction and outcome. The booklet in which the movement's leader reports on its activities bears witness to the opposition and sabotage which his group encountered from the Kanbis, and also to its inability to change the condition of the Dubla agricultural laborers for the better.[4]

Local variations aside, the situation that exists in Chikhligam and Gandevigam is unquestionably not peculiar to these two villages. I conclude from this that my findings there are indicative of the pattern of relationships between landlords and agricultural laborers that has developed throughout south Gujarat.

4. Dave, 1946.

Chapter 11

Postscript. Polarization: More of the "Same" for Everyone

11.1 The Villages Revisited

In the summer of 1971, when I returned to south Gujarat, my primary concern was not the village situation, but I used the opportunity to make several trips to Chikhligam and Gandevigam. In each village I spent about two weeks in all — not very much, but enough to form an impression of what had happened in the meantime.

What struck me first was the absence of visible change. As before, Chikhligam could be reached only by a rough country road, and the bus services, though greatly extended in recent years, still do not reach there. Nor has the village been supplied with electricity. Gandevigam, which already had more public facilities seven years ago, has retained its lead. My impression of greater prosperity in Gandevigam was confirmed when I saw that this village in the garden region had already recovered from the consequences of a flood some years ago. There were few traces of the havoc caused in 1968, when the Ambika River inundated the riverside villages after heavy rainfall in the monsoon.

The population of the district has greatly increased, although growth shows a differential pattern. The birth rate among the higher castes was already relatively low before sterilization came into use as the chief means of reducing it. Among the members of the lower castes, however, restriction of the number of children was much less practiced. The government is pursuing in earnest its plans to curb the rapid population growth, and the organization charged with the implementation of these plans has taken several steps, including the enlistment of the local authorities. The first time I returned to Chikhligam the village headman kept away, because he thought he recognized the car of the district family-planning unit in the jeep that brought me. He was afraid he was once again going to be pestered about the number of applications for sterilization made by men in the reproductive age range. Propaganda to this end is conducted by a female staff member of the family-planning organization, a Dhodia girl recently stationed in the village, who directs her efforts almost exclusively to the members of the lower castes. It is the latter who are most susceptible to the promised bonus which, in specially conducted campaigns, amounts to 80 rupees, a sum that for an agricultural laborer is more than twice his monthly wage. It seems, however, that the possibility of greater results is not the only reason for singling out the tribal castes, such as the Dublas and Dhodias. Among members of the higher castes as well as in government circles the opinion is often voiced that the rapid numerical increase of the lower castes not only affects economic development adversely but also lowers the quality of the population — a dubious argument. However, it is clear that, as a result of population growth, the pressure on the agrarian basis of existence has exceeded the tolerance limit. For those categories that own disproportionately small amounts of land or none at all, intensification of production merely offers a short-term solution.

Because there are few possibilities of finding work outside agriculture and outside the village, the problem is aggravated. Economically, the small district towns in particular are characterized by stagnation rather than growth. Employment in the urban centers has increased very little, so that it can scarcely acommodate the existing local labor supply, let alone absorb the large surplus from the countryside.

11.2 Changes in the Pattern of Agricultural Activities

No drastic changes have taken place in the crop system of either village. In Chikhligam, mangoes and rice have remained the principal market products. Several Dhodia farmers recently turned to chillie cultivation. For this purpose, new wells were sunk and, in some cases, oil pumps installed. In Gandevigam, rice cultivation has greatly increased. With a further extension of the orchard area in this village the shift from mango to *chiku* cultivation — even more profitable and less risky — has continued. The Koli landowners have concentrated more on the cultivation of vegetables and chillies. The increased yields of the crops are not due to any sudden breakthrough in a stagnating agricultural economy, nor does it proceed with such rapidity in south Gujarat that the use of the term "green revolution" is justified. Nevertheless, the New Agrarian Strategy launched by the government in the late nineteen-sixties has undoubtedly resulted in a changeover to new agricultural methods. The use of fertilizers and insecticides is becoming general, and in rice cultivation the introduction of higher-grade seeds increases steadily.

This evolution has, however, only widened the gap between the two villages. A large part of the fertile soil in Gandevigam is permanently irrigated by water from the canals and wells. The fields are scenes of great activity almost the year round. "Progressive" farmers in this garden region achieve two rice crops a year, some of them even three. The intensification of agriculture has increased employment, but mechanization has also progressed. Tractors have come into general use, and these vehicles are also employed for the transport of agricultural products. During the rush season the roads in this region are crowded with traffic.

In comparison, agriculture in Chikhligam is still labor-extensive. Although an irrigation canal crosses the village area, the land lies too high and is, moreover, to uneven to allow irrigation. At present the Land Development Bank of the district is preparing a program of leveling of land which, combined with the use of electricity, could facilitate irrigation in a larger area. It is not easy, however, for Chikhligam to make up the deficiency. The extension of employment has not kept pace with the growth of the population in general and of the agricultural population in particular, while the distribution of agricultural activities over the year is also much more irregular than in Gandevigam.

Differentiation has increased both regionally and socially. It is the large farmers that have profited by the support given by the government to promote production. As they control the cooperative movement and have great influence on the local bureaucracy they are generally the ones to benefit by the facilities offered in this way. The *gram sevak*, whose task as a local-level worker has been reduced from community development in a wide sense to agricultural extension work only, directs his efforts almost exclusively to members of the dominant caste in both villages. Kolis and Dhodias rarely qualify for any credit granted by the government, such as for land improvement. Insofar as they are able to buy better seed and fertilizer for rice cultivation most of them use the crop for household consumption. At any rate this makes it unnecessary for them to buy grain for their daily needs at steadily rising prices in the village shops, as the agricultural laborers have to do. The contribution of the small farmers to production for market consists chiefly of a few garden crops. These labor-intensive products form the principal source of their money income from agriculture and enable them to find employment within their own households. In this way they can avoid the dependence associated with the performance of wage labor for large farmers. In a situation of enlarged socioeconomic inequality, the small landowners strive more or less successfully to retain their independence. They are not systematically included in the development programs, and the advantages that some of them obtain from these programs are derivative and limited in nature.

11.3 The Political Framework

In both Chikhligam and Gandevigam the hegemony of the Anavils is undiminished. In the village council as well as in the local cooperative they are still in control. Various facilities offered by the government are distributed through these institutions, and the mediating role of those in charge enables them to perpetuate their local power. Any request made by a Dhodia from Chikhligam for a loan towards buying an oil pump, for instance, depends on the intercession of the village headman and that of the secretary of the cooperative, both Anavils. The weak economic condition of the small farmers contributes to the political ascendancy of a small elite, and the New Agrarian Strategy has not altered this in any way. Fertilizer, for example, can be bought from the cooperative at prices fixed by the government, but only when paid for in cash. Several Dhodias who cannot do so are supplied with small quantities on credit by a few prominent Anavils at higher prices or in exchange for support in local affairs.

In consequence of the administrative and political decentralization in recent years, which has shifted the center of authority to district level, key positions within the village have become more important than before because of the weight they carry outside it. The village headman, for instance, occupies a seat in the regional council in that capacity. Because he is a member of one or more council committees that allocate government money for such purposes as public works, health care, and education, he is able to obtain considerable facilities, either for the village or for his own benefit. The two Anavils of Chikhligam and Gandevigam meet their fellow caste members or other locally influential people in the council, and the various benefits are distributed within this group on a basis of mutual favoritism. Aside from this they have easy access to the regional bureaucracy. The present village headman of Chikhligam prides himself on his familiarity with some prominent district officials and reinforces his leading position by occasionally also letting non-Anavils profit by his good contacts. Some years ago his cousin, who is now head of the village cooperative, gained his influential position of secretary of the *taluka* branch of the Congress Party, once a stronghold of the dominant caste in south Gujarat. After the split in the top ranks of the party, a large majority of the party supporters continued to follow Morarji Desai, a man of the region. Much the same thing happened in other parts of this conservative state. But their support rapidly crumbled away as the popularity of Indira Gandhi increased, reinforced by her slogans to ban poverty and the successful end of the war against Pakistan. In the months preceding the election of March 1972, many local cadre members joined the New Congress Party. It was not so much ideological motives that made them change their allegiance, but rather fear of losing access to the external sources of power. As already noted, local leadership is based to a large extent on the distribution of facilities provided by government bodies. By changing his allegiance in time, the district branch secretary from Chikhligam was able to avoid being excluded from benefits after the election.

It would not be correct to conclude from such occurrences that the old political system has remained completely unchanged. Notably, the ruling party can no longer count on the influence of the largest landowners in obtaining the votes of the small farmers and the agricultural laborers. This mechanism, based on local patronage, has been replaced by direct mobilization of the weak but numerically dominant groups by organizers from the regional power centers. Nevertheless the old elite has managed to accommodate itself anew to a governing party and therefore has lost little of its authority.

11.4 The Condition of the Intermediate Castes

Whereas the top group has generally succeeded in consolidating its privileged position, most of the Dublas are just as badly off as they were some years ago, and perhaps even more so. Although this account is primarily intended to establish the present relationship between these two categories, I shall begin by giving a brief description of the condition of the Kolis and Dhodias who are between the two strata in Gandevigam and Chikhligam, respectively.

The generally accepted view is that, in the district as a whole, these castes have been able to improve their position. Those who have found work outside agriculture have been especially successful. The skilled employees of the factories and workshops in the small towns of the district are usually members of these castes. A greatly increased interest in education has played an important role in their social rise. With a college education, Kolis and now also Dhodias increasingly succeed in obtaining clerical jobs in government or industry formerly reserved for members of the higher castes. Participation in education is facilitated by the circumstance that many of the schoolteachers, particularly those in the countryside, belong to these intermediate castes. The Kolis and Dhodias have gained political influence as well. By appealing to collective loyalty, candidates of these numerically influential castes were elected to various important posts. Another dimension of the improvement in their position is expressed by urban-based sectarian movements — such as the Moksh Marg, which is active among the Dhodias — urging sanskritization of their style of living.

Does this mean that the Kolis and Dhodias have shaken off the dominance of the Anavils in village life? Any general statement concerning either of these groups would be premature, as a process of differentiation is taking place precisely in the intermediate castes.

As to the Kolis of Gandevigam, some of them have succeeded, partly owing to the Tenancy Acts, in rising to the position of medium-sized landowners. In contrast to the majority of their fellow caste members, who work exclusively with their own relatives, they call on Dublas as additional labor for longer periods. Whenever possible they are inclined to avoid manual labor, if not for themselves, at least for their wives and children, in imitation of the Anavils. These Kolis are better housed (in houses with stone walls and tiled roofs), own more capital goods (especially agricultural implements), and further distinguish themselves by a higher level of consumption. Together with the artisans they form an elite within their caste. Children from such successful Koli families are most likely to take secondary or higher education; after completion of their studies, jobs are found (sometimes being bought) for them in a factory or government office in one of the neighboring towns. On the other hand, the situation of most of the other members of this caste has changed little. Their holdings are small, and they are still typically thrifty and hardworking peasants, in whose households all the members have to take part in the work on the land. They have perceptibly remained behind in income and status. In families with many children, above all, the prospects for the new generation are extremely gloomy.

Among the Dhodias of Chikhligam, who, taken as a whole, are below the level of the Kolis in all respects, this differentiation is even more pronounced. Here, too, the bettered position of a number of households is indicated by, for instance, tiled roofs and the possession of bicycles and transistor radios. During my stay in the village, preparations were being made for the building of the first brick house in one of the Dhodia quarters. Even less than in Gandevigam do the widening economic gaps within this caste reflect disparate incomes from agriculture. Most successful are the labor-recruiters who locally enlist labor gangs for the brickyards. One of them — the most successful — contracts about 500 laborers a year. Together with a brother he owns two trucks, his younger brother is enrolled in the university at Surat, and he himself bought a mango orchard from an Anavil. In short, he is a man of substance. Through the influence of the recruiters the slightly better-paid jobs in the brickyards — supervising labor gangs, driving trucks — usually fall to their fellow caste members. The boys from these financially more secure Dhodia households no longer participate in the yearly migration to Bombay. Several have been trained for higher-qualified jobs and are now employed as schoolteachers, factory hands, or diamond cutters. This top category in particular profits by the special facilities for tribal caste members. The general impression of increased power and independence is above all derived from this minority.

In speaking of ascending castes in a general sense, therefore, insufficient account is taken of the growing differences among Kolis and Dhodias. Those that have lagged behind are not concerned with accentuating the widening distance between them and their successful fellow caste members, while the latter hope to improve their chances of competing with the high castes for economic advantages and political influence by appealing to group solidarity. In the intermediate castes, the process of socioeconomic differentiation remains veiled, so to speak, by the personal interests of both those ascending and those remaining behind.

The continuing polarization in the rural regions of south Gujarat may be most adequately demonstrated by a comparison of the situation in which Anavils and Dublas in Chikhligam and Gandevigam find themselves.

11.5 The Anavils

The Anavils have not lost the stamp of a rural caste. For most members, agriculture is still the principal basis of existence. To be sure, secondary education has become a matter of course for the younger generation, but the possibilities of nonagrarian employment have not much increased, so that migration to towns in or outside the district is impeded. A long period of unemployment after completion of college is the rule rather than an exception, and many find themselves finally forced to accept jobs far below their level of training. There is sharp competition for the few places available in government offices or industry, and candidates from ascending Koli and Dhodia households also seek those jobs in increasing numbers. Although it is true that Anavils, through the influence of fellow caste members in established positions, have succeeded in maintaining their social lead, the remuneration of those urban jobs for which they are qualified is far below the income they can derive from farming. In addition, the influx of members of lower castes has diminished the prestige of the various occupations in which the members of the dominant castes used to be strongly represented — teacher, government employee, bank clerk, and the like.

Urban orientation among the higher castes has, finally, decreased because public facilities in the countryside have improved. Electricity and tap water supply have made life more comfortable in many villages. Bus services have been extended and their frequency increased, so that the villages are less isolated. It should be borne in mind, however, that these facilities benefit, in the main, the rural elite. Their way of life differs little from that of the upper urban class.

The Anavils generally have done very well for themselves. The extent to which they have improved their position is shown, for instance, in the successful development of the regional cooperative of Gandevigam and neighboring villages. Whereas this institution — still primarily an instrument of the dominant caste for their own interests — listed sales totaling about 750,000 rupees in 1962, they had risen to 2,500,000 rupees for the year 1971. Membership increased from 700 to over 900, but cash reserves trebled in the same period, and now amount to nearly 1,000,000 rupees. The price of *bagayat* land, mostly owned by Anavils, has risen to an average of 25,000 rupees an acre in Gandevigam, that of *kyari* to 10,000 rupees or higher. One informant has given up his teaching post. As a result of more careful management of his farm, he now has a substantial income: his medium-sized orchard alone provides him with a yearly revenue of over 9,000 rupees, more than twice the salary he earned as a teacher.

The Anavils of the garden region are not the only ones who have profited by this favorable trend. While the exceedingly fertile soil in Gandevigam yields more per acre, the Anavils in Chikhligam possess, on an average, more land. This does not mean, of course, that there are no differences in prosperity among them. The richest members of the caste had acquired new durable goods in addition to those I noted seven years earlier. Their owning a motorcycle or even an automobile is becoming less unusual. Their houses have been improved, for instance, by tiling on the walls and floors. One Anavil from Gandevigam told me about a holiday tour he had recently made through India which cost him a sum that would have provided a living for two agricultural laborers for a whole year. Following a more traditional spending pattern, another member of the caste recently invited some 1,500 guests on the occasion of the sacred-thread ceremony of his son's son. It is true that only a minority can afford such luxuries, but in general the situation of the Anavil households is one of increased prosperity, and I met with no examples of real deterioration among them. Practically all the Anavils have succeeded in realizing the aim of not performing agricultural labor themselves, and this has further lessened the social distance between Bhathelas and Desais. The sociocultural homogeneity within the dominant caste is reinforced by their strongly developed sense of being the backbone of the social system.

Awareness of their common interest does not, however, preclude clashes. In Chikhligam, for instance, a conflict has arisen between the former village headman and his cousin, the former secretary of the cooperative. They have now exchanged functions, but behind this change — not sensational in itself — there is a fierce struggle for power. Both men strive to be acknowledged as the village leader, a condition for influence outside the village. For this kind of political competition, it would not be correct to use the term factionalism, which would imply the existence of alliances within the local system that are kept up by means of patronage-like relationships. In this sense the Anavils are no longer split into different camps, each with its own superiors. There is, rather, progressive individualization, in which members of the dominant caste, as businesslike farmers, keep a sharp watch on each other and regularly exchange allegations of improper practices in using the canal water, hiring agricultural laborers, and the like. Among the various households, suspicion and jealousy prevail.

Sincere collaboration is therefore still rare, yet there is increasing awareness of the fact that mutual opposition at any rate damages the common interest, for the Anavils feel more threatened in their privileged position than ever before. A feeling for the political scene has, it is true, induced many leaders to pronounce for the N.C.P. at the right moment, but in this they have been followed by only a minority of their conservatively thinking fellow caste members. The rural elite has little use for the socialist course to which the new party has ideologically committed itself. Most Anavils feel shocked by the success of the populist strategy of the N.C.P. They explain the striking victory of Indira Gandhi in the spring of 1972 as a transition to a more left-wing rule — a shift that does not hold much promise for them. They are, however, somewhat reassured by their conviction that the *garibi hatao* program will fail once again. The fact that the regional leadership of the N.C.P. consists predominantly of Anavils and members of other high castes assures continuity and presents a possibility of more or less legally invalidating any too drastic proposals — for instance, the land reform proclaimed by the government.

In the latest election, most members of the dominant caste remained faithful to the Old Congress Party. The young people in particular, however, feel attracted to the Jan Sangh. This militant urban-based right-wing party has recently won many members among the sanskritizing castes in the countryside. A number of young Anavils and Kolis from Gandevigam — disappointed, like so many of their age group, in their expectations of getting good jobs after completing their studies — have joined a newly established branch, in the *taluka* town, of the R.S.S., the paramilitary organization affiliated with this orthodox Hindu party.

The members of the upper social stratum deeply regret the fact that, in the present political system, strength is measured in numbers. They see it as the writing on the wall that the N.C.P. owed its victory to massive support from the lower castes. In the election campaigns, they complain, these categories were incited to offer resistance against the rule of the higher castes and were lured with false promises or money. A widespread allegation against the politicians is that they bought votes of the population in poor neighborhoods through trusted agents.

From the height of their elite position the Anavils give vent to a deep aversion to the democratic system in which they feel outvoted by an illiterate, irresponsible, and inert mass. In their opinion, participation in the elections should be reserved for those who are qualified to judge because of their parentage and education. It is a view in which the superiority of the high castes is asserted and the inferiority of the low castes emphasized.

11.6 The Dublas

My first walk through the Dubla quarter on the fringe of Chikhligam showed me that their material situation had not improved. Many huts were in a state of collapse and offered little or no protection against the winter cold and the monsoon rains. The average number of occupants per hut has greatly increased, and the available space makes it impossible for more and more households to shelter all their members at the same time. Although the hamlet is surrounded by a great deal of waste land belonging to the village, any applications for permission to build new huts there, transmitted to the district authorities with a negative recommendation by the village headman, have remained unanswered.

In Gandevigam it was the Dubla quarter, inundated even in normal rainfall, that suffered most in the flood of some years ago.

It is true that the total damages in this quarter were relatively low, but that was because of the low value of the huts and household effects. The little that the occupants possessed was mostly lost. The aid offered from outside was distributed by the local elite and, if only for that reason, passed the Dublas by, or benefited only the confidants of some of the prominent Anavils. Government money was used to build fifty-two new huts with tiled roofs, but not in the old place. Because the village authorities refused to put higher-lying ground at the Dublas' disposal, a large part of them had to move to a distant site on the territory of a neighboring village. One water well is the only public facility there. Electric power wires pass right over the rows of huts, but there is no connection to them anywhere. Fraudulent economizing in material led to jerry-building. The Anavil secretary of the regional cooperative had to resign when it became publicly known that he had sold wood intended for these huts and had pocketed the proceeds. Lastly, the new Dubla quarter is difficult of access. The men and women now have to traverse much greater distances to reach the houses and fields of the farmers, which creates much hardship especially during the rainy season.

The deplorable condition of the Dublas is, of course, closely connected with the lack of sufficient and continuous employment. From Chikhligam, the yearly migration to the brickyards goes on as before, because of need but not by choice. The continual alternation of work situations, which often entails the breaking up of households for long periods, makes any stable existence impossible. It also means complete subjection to a labor system in which the employer, owing to the temporary nature of the engagements, rejects any obligation to provide for the most elementary needs of his employees. This sometimes leads to their being treated inhumanely. During my stay, for instance, a grandfather, a father, and two daughters returned to the village in mid-season. They suffered from typhoid fever and had been sent away from the brickyard so as not to infect the other laborers. Gravely ill and penniless, they had found their way back, traveling surreptitiously by train and getting a lift in an oxcart. Once they were in Chikhligam, not one of the farmers for whom members of this household sometimes worked gave them any aid. I found these people in their hut, uncared for and without food, lying on jute bags. Two of them died of the fever the next day.

The Dublas of Chikhligam are dependent on their income from agriculture for at least part of the year, and most of their fellow caste members in Gandevigam depend on it all of the time. During recent years, the living and working conditions of agricultural laborers in India have aroused increasing interest, particularly as to how the New Agrarian Strategy has affected them. The information that is gradually coming to light shows that the situation of this numerous category of the population has not noticeably improved, and in many cases has deteriorated. This seems to be true of the Dublas as well. Although in both villages agricultural laborers' wages have practically doubled as compared with seven years ago — farm servants now earn 1 to 1.5 rupees and day-laborers 1.5 to 2 rupees — the prices of basic living needs have risen correspondingly or even more, so that a decent existence is still impossible. Employment in Chikhligam has not greatly increased, although in Gandevigam the demand for casual laborers to work for farmers with large and medium-size holdings has risen. Even in the labor-intensive crop pattern of Gandevigam, however, the labor supply shows a much greater rate of increase. This impression of the situation in the two villages is confirmed by the evolution of the wage level and of employment in the whole of south Gujarat. On the basis of 1962–63 equal 100, the input of labor in 1967–68 had risen to 114, but this was offset by a fall in the real wage level to 78 in the same period.[1]

These data, though illustrative in themselves, reflect a more fundamental transformation. They indicate an alteration in the nature of rural organization — a changeover to a capitalist mode of production. The commercialization of the relationship between farmers and agricultural laborers is reflected in a number of labor forms which have become increasingly conspicuous during the past decade. The percentage of servants has decreased but will probably remain at the present level in view of the need among the large farmers of a small fixed core of labor. For attached labor, employers prefer Dublas who have some skills, can handle the new tools, and show an insight into the new agricultural techniques. A submissive attitude is another requirement. Agricultural laborers who are known to be "troublesome" and "pugnacious" are rejected, insofar as these more militant Dublas have not already renounced

1. P. Bardan, "Green Revolution and Agricultural Labourers," in *Economic and Political Weekly*, vol. V, no. 29–31, July 1970.

servitude themselves. Casual laborers are employed on a contractual basis, and the landowners can usually draw on an abundant supply. For the harvest of some products, for instance chillies and cotton, piece-work has been introduced on a large scale, a form of labor in which children from agricultural laborers' households are employed more than before. Thirdly, contract labor has gained ground. Labor gangs undertake to carry out a specific amount of work — harvesting rice, grass, or mangoes — at a prearranged price per day or land unit.

The most drastic change in the labor system is undoubtedly the mobilization of agricultural labor across considerable distances. Although Dublas from many villages in south Gujarat have been accustomed for a generation to migrate to the brickyards and saltpans every year, agricultural work is now increasingly being performed by laborers from elsewhere. This migratory labor has assumed enormous proportions. For the harvest of sugar cane, for instance, the cooperative factories with which the bigger landowners are associated recruit thousands of laborers every year in the western districts of Maharashtra. The workers are small peasants and landless laborers, who have no means of existence during the dry season in their home area. With or without their oxcarts they come to south Gujarat for a period of five or six months each year, traveling from village to village cutting cane and transporting it to the factories. In the same way, individual owners of mango orchards and grass fields or dealers in these products contract laborers in distant villages for the harvest period through the agency of labor-recruiters. The labor supply is less and less determined by local conditions. Large-scale migratory labor has a wage-depressive effect and facilitates replacement in case of insufficient performance or too drastic demands. The abundant supply combines with the enlargement of scale to foster competition, and labor comes on the market at a price far below the minimum level of existence.

Yet the economic advantage gained by the farmers is not the only factor that has led to this state of things. The new forms of labor enable the Anavils to limit to a large extent their dependence on the performance of the Dublas, a source of boundless irritation to them. Accelerated changeover to mechanization of the agricultural production, even when clearly not profitable from the viewpoint of efficient management, reflects the desire of the farmers to do away as much as possible with their need for agricultural laborers. Another solution for them is to keep their relations with these subordinates as distant and businesslike as possible by leaving them to intermediaries. The large farmers appoint overseers, charge servants with the negotiations to be conducted with casual laborers, or deal exclusively with gang-foremen or recruiters. The latter receive advance money, arrange the distribution of wages on pay-day and hold themselves responsible for the completion of the arrangements. Conversely, these mediating roles provide an answer to the growing aversion among agricultural laborers to binding themselves individually to landowners. Extending credit to Dublas is increasingly done through *mukadams*, who in this way provide themselves with a pool of labor which they can utilize as it is needed.

The loose and businesslike ties with the farmers which manifest themselves in all these labor forms have led to a further hardening of the relationship on both sides. In the eyes of many Anavils, Dublas are good-for-nothing *kamchor*, labor thieves. Not all landlords, it is true, act with the same harshness, but as a Dubla from Gandevigam put it, "One Anavil kicks you from his field when you have gathered wood without permission, another tacitly allows it; but both of them refuse actual help." Fearing the destruction or theft of crops, more and more landowners have decided to fence in their property with barbed wire, as well as to have a closer watch kept on the fields. The polarization between, on the one hand, the larger farmers who do not personally work, and, on the other, the landless laborers rests on the fact that the former no longer recognize the right of the latter to share and participate in agrarian production. I pointed out these indications of a large-scale rural pauperization when I closed my previous investigation. What is, therefore, quite frequently regarded as a negative consequence of the green revolution is, in my view, actually the condition on which this process started. Partly as a result of government policy the contrasts in rural regions between the weakest and the strongest party have sharpened. The term "green revolution" still applies, but not, as I see it, primarily to a sudden spurt in farm production owing to a technological breakthrough. More important is the fundamental shift in the mode of production to capitalism as the base of agriculture.

The economic situation of the Dublas, miserable as it was, has deteriorated even more. As landless laborers for whom work has become a godsend even when it does not bring in enough to make ends meet, they have been forced further into a corner. Providing for a household requires continuous improvisation.

Besides their exploitation, another aspect of the poverty of the Dublas is their increasing isolation. Not only has the landlords' traditional help and protection disappeared, but they do their utmost to obstruct any direct contact of their former clients with external sources of power. By taking advantage of the ignorance and powerlessness of the Dublas, some Anavils use their influence and their familiarity with laws and regulations to prevent outside support for Dublas. The Dublas in both villages complain of the opposition they encounter in their attempts to qualify for the facilities which the government expressly makes available to the tribal castes. Anavils and members of other high castes, they say bitterly, bar the way for us everywhere. Those few Dublas who have heard of the *garibi hatao* program have no illusions whatsoever that its generous promises will ever be fulfilled. That the social isolation in which the Dublas find themselves has increased is strikingly demonstrated by their removal from the residential nucleus in both Gandevigam and Chikhligam. Regrouped on the fringes of these and many other villages they may rightly be called the new untouchables of south Gujarat.

From various parts of India there have recently been reports of agricultural laborers who are beginning to offer resistance against the exploitation to which they are exposed. It is rightly pointed out, however, that if growing economic inequality is a prerequisite for political unrest, it is nonetheless not sufficient in itself.[2] Large-scale protest movements in reaction against a deteriorated position in the production system need to be organized, but in a situation of pauperization the chances of joint action are drastically limited.

The horizontal ties among the Dublas have remained weak. Some are more oppressed than others. As sharecroppers, servants, casual laborers, and gang laborers they have no parallel interests. The limited favors that can be obtained in a situation of extreme and continuing scarcity lead to patronage-like relationships. An attitude of protest, for instance, is incompatible with the need of showing oneself worthy of consideration by behaving "well." Intra-class antagonism manifests itself in dissension and mutual watchful-

2. T. K. Oommen, "Agrarian Tension in a Kerala District: An Analysis," *I.J.I.R.* Reprint Series no. 15, n.d.

ness. If someone refuses to work for 1 rupee, another is always prepared to take his place, as I was told time and time again. The Dublas are aware of their isolation and attribute it to a lack of leaders of their own, "who can show the way." But who will venture to risk standing up for others against the powerful Anavils, and who, for that matter, expects to get a hearing from the arrogant and unwilling bureaucrats? In the situation of poverty itself there are undoubtedly elements that constitute obstacles to mutual solidarity. Incompetence and ignorance, often considered typical of pauperized groups, have also rendered the members of the landless castes helpless and passive in their dealings with outsiders. On the other hand it is the aggregate of inequality — economic and political — along with social polarization that forms the basis for a growing collective perception of injustice, even if fear of sanctions usually limits any expression of it to verbal avowals during encounters in their own hamlets. Indeed, their awareness of unity is further stimulated from outside because Anavils regard all the Dublas as being cut from the same cloth. The landlords are strikingly ill-informed about anything concerning their laborers, including the contrasts among them, and what is more, they are totally indifferent to them.

One organization that is active among the Dublas is the *halpati seva sangh*, stemming from the Surat district, which is still its main field of action (see chapter 10). In recent years it has extended its activities to the Bulsar district. With financial support from the government it has opened small offices in nearly all *taluka* towns as bases for one or two staff members who try to get through to the Dublas in the countryside. Social workers and local politicians form the nucleus of this voluntary movement, which aims at improving the social position of this landless caste. Among the efforts to that end, education occupies an important place. In the Dubla quarters of an increasing number of villages the organization's members have established kindergartens, with free meals for the children during school hours. Some six months ago the *halpati seva sangh* opened a primary boarding school in Chikhligam, but despite the efforts of the local staff, school attendance remains low. Pupils of the two institutions of secondary education that the organization has started elsewhere in the district belong in the majority to other castes, for lack of Dubla applicants.

As in the other activities undertaken by the *halpati seva sangh*, the education they offer aims at adjustment rather than awareness. Alcoholic excess is combated mainly on moral grounds. Groups may apply for musical instruments so as to hold *bhajan* gatherings, in which religious songs extolling Hindu values are sung. To keep the Dublas from consulting the *bhagat*, their own medicine man, chests containing medicines have been distributed among the villages. Most components of the program are on this sociocultural level. The movement has little to offer in the way of measures towards improvement of the Dublas' economic condition. By paying the expenses of artisan training, the organization tries to provide an alternative to the forlorn existence of the agricultural laborers, but only a limited number of young people can profit by it. Still awaiting implementation is a plan to establish some labor cooperatives whose members would work for farmers on a contractual basis and, in the slack period, would contract for the construction and maintenance of local roads and irrigation canals.

The *halpati seva sangh* can best be characterized as a movement aiming at adapting the Dublas to the existing social system on the most favorable conditions possible. The movement derives from Gandhian principles of gradualism and conciliation of contrasts. In this view, any schooling to promote class consciousness and class struggle is not so much tactically injudicious as sinful and therefore objectionable.

At present a more radical approach would, it appears, have few chances of success. In south Gujarat, the Land Grab Movement, launched in the autumn of 1970 by left-wing parties, ended in failure. The agitation conducted by a bourgeois leadership from the district towns only impelled the Dublas of a few hamlets to take action. The risks attached to occupation of the land owned by large farmers were too great. In some villages near Chikhligam, where the action was led by the C.P.I. — a party with only a handful of members in the whole Bulsar district — the laborers were taught their place. A strong police force dispersed the crowd of men, women, and children by brute force. Many participants were taken to distant prisons and the instigators received harsh sentences. The *halpati seva sangh* refused to help any of those that had joined. The landlords went round the laborers' hamlets to say that this had been a warning and that any further action would be put down with even greater severity. Although such experiences may intensify the existing awareness of oppression, they do not necessarily lead to more vehement confrontation.

The *halpati seva sangh* maintains close ties with the government party. Through its staff members the Dublas are urged to vote for the N.C.P. After the split in the Congress movement some years ago, the district authorities of Bulsar — who before the recent election supported the O.C.P. — promptly withheld the subsidy for the *halpati seva sangh.* The split in the formerly united party has brought some clarity, inasmuch as the political choice of the upper stratum at least no longer coincides with that of the lower classes. This is not to say that the interests of the landless laborers are now better protected. It seems probable that the Dubla hamlets have become the vote banks of the regional leaders of the N.C.P., most of whom derive from the higher social strata. Members of the landowning castes are amply represented among the leaders of the *halpati seva sangh* — indeed, they are headed by an Anavil, who in his attitude and actions corresponds perfectly to the traditional image of the *dhaniamo.* Small wonder, therefore, that the Dublas talk with cynicism about this organization. They have learned to distrust profoundly the promises made by anyone. They regard the *halpati seva sangh* members who function as brokers and mobilizers as agents of the upper castes, but this is not always true. The local cadres, mostly consisting of young and skilled Dublas, admit to being not much more than pawns and only dare to voice their criticism in whispers for fear of dismissal. In their thought, however, they show themselves much more radical, and a change of course might be possible if the organization were taken over by them.

In this analysis I have emphasized the obstacles in the path of those trying to create organized solidarity. Intraclass tensions play a role, and nonantagonistic mobilization of the agricutural laborers conceals fundamental contradictions. It would nonetheless be incorrect to suppose that awareness of unity still needs to be awakened. Closer contact shows that class feelings are certainly present, but under the existing power structure it would be imprudent and injudicious for the laborers to display them too openly. The exploitation is heavy and the laborers' situation is confused. Complete bondage to a single landlord has for most of them been replaced by partial dependence on farmers, shopkeepers, labor-recruiters, government functionaries, and politicians presenting themselves as social workers. But the different forms of oppression are often very sharply distinguished and brought under a common denominator by those who suffer from them.

The situation does not offer much hope for the near future. Escalation of violence seems almost unavoidable. Despite the striking moderation of its demands, the *halpati seva sangh* remains a thorn in the side of the farmers. Shortly after my departure a local leader of the organization — a Dubla whose father fled from Gandevigam years ago after a quarrel with his landlord — was murderously attacked by a field watchman under orders of an Anavil. The Dublas, on the other hand, are losing their fear precisely as a result of pauperization. When I was taking leave, one of them said that, before my next visit, I would have heard of large-scale open clashes between farmers and agricultural laborers.

Glossary

ādivāsī	collective name of tribal groups and castes
ādu	dried ginger
Ahīr	caste of shepherds
Anāvil Brahman	dominant caste of landlords in south Gujarat
āshram	institution of social work on Gandhian lines
āvat	term for *jajmāni* system current in south Gujarāt
bāval	a tree, *acacia arabica*
bāgāyat	garden land
Baniā	caste of Merchants
bhātā	farm servant's daily ration in grain
Bhāṭhelā	section within the Anāvil caste
bīḍī	native cigarette
chākar	farm servant, at the same time house servant
chhās	skimmed milk, residue of the preparation of *ghī*
chīku	a fruit, *sapodilla*
dāl	lentils
Darjī	caste of Tailors
Deśāī	section within the Anāvil caste; originally, designation of local power elite and revenue farmers; derived from *dēsh* (land)
deśāīgīrī	government compensation for lost Deśāī rights
dhaṇiamo	traditional name for master of farm servant
dharmashālā	guest house
Dheḍ	caste of Weavers
Dhodia	a tribal caste
dīvālī	Feast of Lights
Dublā	a tribal caste; also called *halpati*
gām	village
Gāmit	a tribal caste
Gāruḍā	priest caste of Untouchables
Ghānchī	caste of Oil-pressers, at present often shopkeepers
ghī	"clarified" butter
govālio	cowherd
gulām	slave

guṛ	roughly pressed and evaporated cane sugar
hal	plow
haladar	species of ginger
hālī	designation of farm servant who was traditionally bound to a landlord
hālīprathā	pattern of relationships in which the traditional relationship between landlords and farm servants was institutionalized
harekwālī	maid
Harijan	collective name for castes of Untouchables; lit., people of God
holī	spring festival
ināmdār	functionary whose office entitled him to a land grant or to part of the land rent
jarāyat	high-level, dry land
juvār	species of millet
kacchā	not pure, inferior
kālīparaj	description of tribal groups, non-Hindus
Kaṇbī Paṭēl	dominant caste of landowners in the northern part of the plain of south Gujarat
khāvati	advance payment in money or in kind on work to be performed
kodrā	inferior species of grain
Kolī	caste of small farmers in the plain of south Gujarat; along the coast also fishermen
kyārī	low-lying land, surrounded by embankments, used for rice cultivation
mahāl	administrative unit, smaller than a subdistrict
Mochī	caste of Shoemakers
mukādam	labor recruiter
nāglī	inferior species of grain
Nāyakā	a tribal caste
pālkhī	palanquin
panch	council
panchāyat	village council, caste council
Pārsī	group, descended from the Persians, who settled in south Gujarat in the seventh century
paṭlā	labor gang
paṭēl	village functionary with police authority
Pedivālā	highest status group among the *Deśāīs*

phalia	village quarter or street
phaṇas	jackfruit
pīpar	long pepper (a medicinal herb)
pīvānum	drink money
rājā	lord
roṭlī	pancake made of unleavened dough
riāyatvārī	system of direct land tax collected by the government without intermediaries
rupee	national monetary unit in India; the traditional subdivision into 16 annas is still used by the population in calculating small amounts
sandhal	exchange of labor
sāṛī	women's outer garment
satī-patī	movement among tribal groups in the eastern part of south Gujarāt, which has a religious undertone
seer	unit of weight, about two pounds
sūraṇ	elephant's foot (a tuber)
talāṭī	village clerk
tāluka	subdistrict
ṭopī	men's headgear
ujlīparaj	all Hindu castes without a recent tribal past
Vairāgī	local type of *sādhu*
vakil	solicitor, lawyer
vāl	a leguminous plant
Vāland	caste of Barbers
vāvlā	plot of land which, in the past, was given to a farm servant for his own cultivation

Abbreviations

A.A.	*American Anthropologist*
A.I.	*America Indigena*
B.J.S.	*The British Journal of Sociology*
B.T.L.V.	*Bijdragen tot de Taal-, Land- en Volkenkunde*
C.I.S.	*Contributions to Indian Sociology*
C.S.S.H.	*Comparative Studies in Society and History*
E.A.	*Eastern Anthropologist*
ETHN.	*Ethnology*
E.W.	*Economic Weekly* (continued as *Economic and Political Weekly, E.P.W.*)
H.O.	*Human Organization*
I.E.S.H.R.	*The Indian Economic and Social History Review*
I.J.A.E.	*Indian Journal of Agricultural Economics*
I.J.S.W.	*Indian Journal of Social Work*
I.S.	*Indian Sociologist*
J.A.F.	*Journal of American Folklore*
J.G.R.S.	*Journal of the Gujarat Research Society*
J.U.B.	*Journal of the University of Bombay*
M.I.	*Man in India*
S.B.	*Sociological Bulletin*
S.G.	*Sociologische Gids*
S.R.—N.S.	*Selections from the Records of the Bombay Government, New Series*
S.W.J.A.	*Southwestern Journal of Anthropology*

Bibliography

AGRICULTURAL LABOUR ENQUIRY

1955 *Report on an Intensive Survey of Agricultural Labour, 1950–51*, vol. I (All India) and vol. V (West India), Delhi

AGRICULTURAL LABOUR IN INDIA

1960 *Report on the Second Enquiry, 1956–57*, vol. I (All India) and vol. V (West India), Delhi

AGRICULTURAL LABOUR IN INDIA

1962 Ed. by V.K.R.V. Rao, Bombay

"AGRICULTURAL WAGES AND SYSTEMS OF WAGE PAYMENTS"

1948 In *I.J.A.E.*, vol. III, no. 1, "Proceedings of the Eighth Conference"

BAILEY, F. G.

1957 *Caste and the Economic Frontier*, Manchester

BAINES, J. A.

1912 *Ethnography, Castes and Tribes*, Strassburg

BAKS, C. J., ET AL.

1965 "De betrouwbaarheid van informatie in een kastesamenleving in Gujarat, India" [The reliability of information in a caste society in Gujarat, India], in *S.G.*, vol. 12, 167–74

BAKS, C. J., J. C. BREMAN AND A.T.J. NOOIJ

1966 "Slavery as a System of Production in Tribal Society," in *B.T.L.V.*, vol. 122, 90–109

BAKS, C. J.

1969 *Afschaffing van pacht; een onderzoek naar de gevolgen van de afschaffing van pacht in twee dorpen van Zuid-Gujarat, India* [An investigation into the consequences of the abolishment of tenancy in two villages of south Gujarat, India], Ph.D. thesis, Amsterdam

BARDAN, P.

1969 "Green Revolution and Agricultural Labourers" in *E.P.W.*, vol. V, no. 29–31, July 1970.

BARODA ECONOMIC DEVELOPMENT COMMITTEE, 1918–19

1920 *Report*, Bombay

BARODA ENQUIRY COMMISSION

1873–75 Vol. I, *Correspondence*, Bombay

BARTH, F.

1960 "The System of Social Stratification in Swat, North Pakistan," in E. Leach (ed.), *Aspects of Caste in South India, Ceylon and North-West Pakistan*, Cambridge University Press

BEIDELMAN, T. O.

1959 *A Comparative Analysis of the Jajmani System*, Monographs of the Association for Asian Studies, VIII, New York

BELLASIS, A. F.

1854 "Report on the Southern Districts of the Surat Collectorate," in *S.R. — N.S.*, Bombay

BENNETT, J. W. AND I. ISHINO

1963 *Paternalism in the Japanese Economy; Anthropological Studies in Oyabun-Kobun Patterns*, University of Minnesota Press

BETEILLE, A.

1965 *Caste, Class and Power: Changing Patterns of Stratification in a Tanjore Village*, University of California Press

BOMBAY PRESIDENCY

Miscellaneous Official Publications; Minute 15–10–1830, Malcolm Peyt

BREMAN, J.

1969 "Veranderingen in de betrekkingen tussen landheren en landarbeiders in Zuid-Gujarat, India" [Changes in the relationships between landlords and agricultural laborers in south Gujarat, India], in *S.G.*, vol. XVI, 395–402

BROOMFIELD, R. S. AND R. M. MAXWELL

1929 *Report of the Special Enquiry into the Second Revision Settlement of the Bardoli and Chorasi Talukas*, Bombay

CENSUS, IMPERIAL, OF 1881

1882 Vol. II, Bombay

CENSUS OF INDIA, 1901

1902 Vol. XVIII, Baroda, pt. 1, Bombay

CENSUS OF INDIA, 1911

1911 Vol. XVI, Baroda, pt. 1, Bombay

CENSUS OF INDIA, 1921

1922 Vol. III, Bombay Presidency, pt. 1, Bombay; vol. XVII, Baroda, pt. 1, Bombay

CENSUS OF INDIA, 1931

1932 Vol. XIX, Baroda, pt. 1, Bombay

1933 Vol. VIII, Bombay Presidency, pt. 1, Bombay

CENSUS OF INDIA, 1941

1941 Vol. XVII, Baroda

CENSUS OF INDIA, 1951

1952 *District Census Handbook: Surat District*, Poona

CENSUS OF INDIA, 1961

1962 Paper 1 of 1962, "Final Population Totals," Delhi

CENSUS 1961, GUJARAT

1964 *District Census Handbook XVI: Surat District*, Ahmedabad

CHOKSEY, R. D.

1968 *Economic Life in the Bombay Gujarat (1800–1939)*, London

COHN, B. S.

1955 "The Changing Status of a Depressed Caste," in McKim Marriott (ed.), *Village India: Studies in the Little Community*, The American Anthropological Association Memoir no. 83, 53–77

1959–60 "Some Notes on Law and Change in North India," in *E.D.C.C.*, vol. VIII, 79–93

1960–61 "The Pasts of an Indian Village," in *C.S.S.H.*, vol. III, 241–49

CORRESPONDENCE

Between the Directors of the East India Company and the Company's Government in India on the Subject of Slavery, Account and Papers, vol. XVI, Session 15–11–1837 — 16–8–1838, pt. LI

DAVE, J.

1946 *Halpati-mukit; hālī prathā ane mukitadānni hilachāl*, Ahmedabad

DESAI, A. R. (ED.)

1961 *Rural Sociology in India*, The Indian Society of Agricultural Economics, Bombay

DESAI, B. A.

The Marketing of Mangoes in the Surat District, Ph.D. thesis, Bombay

DESAI, D.

1942 "Agrarian Serfdom in India," in *I.S.*, 1–14

DESAI, G. G.

1906 *Some Experiences of a Mamlatdar-Magistrate's Life*, Ahmedabad

DESAI, I. P.

1964 *The Patterns of Migration and Occupation in a South Gujarat Village*, Poona

DESAI, M. B.

1948 *The Rural Economy of Gujarat*, Bombay

1952 "Rural Economy of Reconstituted Gujarat," in *J.G.R.S.*, vol. XIV, 111–48

DESAI, M. B. AND C. H. SHAH
1951 "Problems of Farm Labour in Gujarat," in *I.J.A.E.*, vol. VI, 4ff.

DESAI, M. H.
1929 *The Story of Bardoli, Being a History of the Bardoli Satyagraha of 1928 and Its Sequel*, Ahmedabad

DEVOS, G.
1966 "Conflict, Dominance and Exploitation in Human Systems of Social Segregation: Some Theoretical Perspectives from the Study of Personality in Culture," in A. de Reuck and J. Knight (eds.), *Conflict in Society*, Ciba Foundation Publication, London, 60–81

DEY, S. K.
1964 *Community Development: A Bird's-Eye View*, London

DONDE, W. B.
1951 *Rural Labour in the Konkan*, Ph.D. thesis, Bombay

DOSABHAI, E.
1894 *A History of Gujarat from the Earliest Period to the Present Time*, Ahmedabad

DOVRING, F.
1965 "Bondage, Tenure, and Progress: Reflections on the Economics of Forced Labor," in *C.S.S.H.*, vol. VII, 309ff.

DUMONT, L.
1966 "The Village Community from Munro to Maine," in *C.I.S.*, vol. IX, 67–89

ELPHINSTONE, MOUNTSTUART
1884 *Selections from the Minutes and Other Official Writings of the Honourable Mountstuart Elphinstone*, by G. N. Forrest, London

ENTHOVEN, R. E.
1920–22 *The Tribes and Castes of Bombay*, 3 vols., Bombay

EPSTEIN, T. S.
1962 *Economic Development and Social Change in South India*, Oxford University Press
1967 "Productive Efficiency and Customary Systems of Rewards in Traditional South India," in M. Lipton (ed.), *Themes in Economic Anthropology*, A.S.A. Monograph Series, no. 6, London

ETHERIDGE, A. T.
1868 *Report on Past Famines in the Bombay Presidency*, Bombay

FRYKENBERG, R. E.
1965 *Guntur District: A History of Local Influence and Central Authority in South India*, Oxford

GAZETTEER OF THE BOMBAY PRESIDENCY
1877 Vol. II, Surat and Broach, Bombay
1882 Vol. XIII, pt. 1, Thana, Bombay
1883 Vol. VII, Baroda, Bombay
1901 Vol. IX, Gujarat Population: Hindus, Bombay

GAZETTEER OF THE BARODA STATE
1923 2 vols., Bombay

GAZETTEER OF INDIA
1962 Gujarat State, Surat District, Ahmedabad

GOUGH, K.
1955 "The Social Structure of a Tanjore Village," in McKim Marriott (ed.), *Village India*, 36–52
1960 "Caste in a Tanjore Village," in E. R. Leach (ed.), *Aspects of Caste in South India, Ceylon and North-West Pakistan*, Cambridge
1960 "Review of T. O. Beidelman, *A Comparative Analysis of the Jajmani System*," in *E.D.C.C.*, vol. IX, 83–91

GOULD, H. A.
1958 "The Hindu Jajmani System: A Case of Economic Particularism," in *S.W.J.A.*, vol. XIV, 428–37
1964 "A Jajmani System of North India: Its Structure, Magnitude and Meaning," in *ETHN.*, vol. III, 12–41

HABIB, I.
1963 *The Agrarian System of Moghul India (1556–1707)*, London

HARPER, E. B.
1959 "Two Systems of Economic Exchange in Village India," in *A.A.*, vol. LVI, 760–78
1968 "Social Consequences of an 'Unsuccessful' Low Caste Movement," in *C.S.S.H.*, suppl. 3, *Social Mobility in the Caste System in India*, 36–65

HISTORICAL SELECTIONS FROM BARODA RECORDS
1955 *New Series*, vol. I, 1826–35, Baroda

HOLMBERG, A. R.
1967 "Algunas Relaciones entre la privación Psico-Biologica y el Cambio Cultural en los Andes," in *A.I.*, vol. 27, 3ff.

HOMMES, E. W.
1970 *Over de evaluatie van plattelands-planning in India* [On the evaluation of rural planning in India], Ph.D. thesis, Amsterdam

HOVE
"Tours for Scientific and Economic Research Made in Gu-

zerat, Kattiawar, and the Conkuns in 1787–8," in *S.R.*, no. 16, *N.S.*

ISHWARAN, K.

1966 *Tradition and Economy in Village India*, London

JAMABANDI

1907 *Revision Settlement Report of the Gandevi Taluka of the Navsari Division, 1906–1907*, Baroda

JOSHI, V. H.

1966 *Economic Development and Social Change in a South Gujarat Village*, The Maharaja Sayajirao University of Baroda

KAPADIA, K. M.

1966 "Industrial Evolution in Navsari," in *S.B.*, vol. XV, 1–24

KEATINGE, G.

1921 *Agricultural Progress in Western India*, London

KISHORE, J.

1924 "The Village Labourer in Western India," in *Hindustan Review*, 425ff.

KLOOSTERBOER, W.

1954 *Onvrije arbeid na de afschaffing van de slavernij* [Unfree labor after the abolishment of slavery], The Hague

KOLENDA, P. M.

1963 "Toward a Model of the Hindu Jajmani System," in *H.O.*, vol. XXII, 11–32

KOSAMBI, D. D.

1956 *An Introduction to the Study of Indian History*, Bombay

KOTOVSKII, G.

1956 "Indian Agricultural Labourer; a Soviet View," in *E.W.*, April 14, 443ff.

KUMAR, D. K.

1965 *Land and Caste in South India: Agricultural Labour in the Madras Presidency during the Nineteenth Century*, Cambridge University Press

LEACH, E. R.

1967 "Caste, Class and Slavery; the Taxonomic Problem," in A. de Reuck and J. Knight (eds.), *Caste and Race*, Ciba Foundation Publication, London, 5–16

"LETTER OF COLLECTOR OF SURAT, 9–4–1823"

In *Bombay Revenue Proceedings, 1823*, Range 368, vol. XXXVIII, 4816

LEWIS, O. AND V. BARNOUW

1967 "Caste and the Jajmani System in a North Indian Village," in J. M. Potter *et al.* (eds.), *Peasant Society, A Reader*, Boston, 110–34

LORENZO, A. M.
1943 *Agricultural Labour Conditions in Northern India*, Bombay
1945 "Agrestic Serfdom in Northern India," in *I.J.S.W.*, vol. V, 133–41

MAYER, A. C.
1958 "The Dominant Caste in a Region of Central India," in *S.W.J.A.*, vol. XIV, 407–27

MEHTA, B. H.
1934 "A Summary of the Economic Life of an Aboriginal Tribe of Gujarat," in *J.U.B.*, vol. II, 311–59

MEHTA, J. M.
1930 *A Study of the Rural Economy of Gujarat*, Baroda

MEHTA, S.
1930 *Marriage and Family Life in Gujarat*, Ph.D. thesis, Bombay

MOORE JR., B.
1967 *Social Origins of Dictatorship and Democracy: Lord and Peasant in the Making of the Modern World* (paperback ed.), Boston

MORELAND, W. H.
1920 *India at the Death of Akbar*, London
1929 *The Agrarian System of Moslem India*, Cambridge

MORRIS, M. D.
1966 "Economic Change and Agriculture in Nineteenth Century India, a Review of Kumar's Study," in *I.E.S.H.R.*, vol II, 185–209

MORRISON'S REPORT
Letter from Collector of Surat dated 13–11–1812

MUKERJEE, R.
1933 *Land Problems of India*, London
—— *The Economic History of India: 1600–1800*, Bombay

MUKHTYAR, G. C.
1930 *Life and Labour in a South Gujarat Village*, Calcutta

NAIK, T. B.
1950 "Songs of the Anavils of Gujarat," in *M.I.*, vol. XXX, 29–56
1953 "Family in Gujarat," in *J.G.R.S.*, vol. XV, 117–38
1957 "Dubla Songs," in *Vanyajati*, 141–49
1957 "Social Status in Gujarat," in *E.A.*, vol. X, 173–82
1958 "Religion of the Anavils of Surat," in *J.A.F.*, vol. LXXI, 389–96
1958 "The Strains of a Social System," in *J.G.R.S.*, vol. XX, S1–S11

NEALE, W. C.

1962 *Economic Change in Rural India: Land Tenure and Reform in Uttar Pradesh, 1800–1955*, Yale University Press

NIEBOER, H. J.

1910 *Slavery as an Industrial System*, The Hague

OOMMEN, T. K.

—— *Agrarian Tension in a Kerala District*, I.J.I.R. Reprint Series no. 15

OPLER, M. AND R. D. SINGH

1948 "The Division of Labor in an Indian Village," in C. S. Coon (ed.), *A Reader in General Anthropology*, New York

ORENSTEIN, H.

1962 "Exploitation of Function in the Interpretation of Jajmani," in *S.W.J.A.*, vol. XVIII, 302–16

1965 *Goan: Conflict and Cohesion in an Indian Village*, Princeton University Press

PANDYA, B. V.

1959 *Striving for Economic Equality*, Bombay

PAPERS . . .

1858 *Relating to a Summary Settlement of Alienated Revenues in the Bombay Presidency*, Revenue Department, Bombay

PAPERS . . .

1895 *Relating to the Settlement of the Hereditary District Officers' Watans in the Deccan and Gujarat*, W. G. Pedder, *S.R.*, no. 174, *N.S.*, Bombay

PAPERS . . .

1897 *Relating to the Revision Survey Settlement of the Chikhli Taluka of the Surat Collectorate*, *S.R.*, no. 381, *N.S.*, Bombay

PAPERS . . .

1897 *Relating to the Revision Survey Settlement of the Bardoli Taluka*, *S.R.*, no. 359, *N.S.*

PAPERS . . .

1900 *Relating to the Revision Survey Settlement of the Bulsar Taluka of the Surat Collectorate*, Bombay

PAPERS . . .

1900 *Relating to the Revision Survey Settlement of the Jalalpor Taluka of the Surat Collectorate*, *S.R.*, no. 350, *N.S.*, Bombay

PATEL, J. M.

1964 "Agricultural Labour in a South Gujarat Village," in V. S. Vyas (ed.), *Agricultural Labour in Four Indian Villages*, Sardar Patel University, Vallabhai Vidyanagar, 92–119

PATEL, S. J.

1952 *Agricultural Labourers in Modern India and Pakistan*, Bombay

POCOCK, D. F.

1957 "The Bases of Faction in Gujarat," in *B.J.S.*, vol. VIII, 295–306

1962 "Notes on *jajmāni* Relationships," in *C.I.S.*, no. 6, 78–95

PRIVILEGES . . .

1955 *Provided by the Bombay State Government for Backward Classes*, Bombay

RAMAKRISHNAN, K. C.

1948 "Systems of Wage Payments in Agriculture with Special Reference to Madras," in *I.J.A.E.*, vol. III, no. 1

REPORT . . .

1951 *of the Congress Agrarian Reforms Committee*, Delhi

REPORT . . .

1900 *of Famine Operations in the Baroda State for the Year 1899–1900*, Baroda

REPORT . . .

1950 *of the Halī Labour Enquiry Committee*, M. L. Dantwala and M. B. Desai, Bombay (not published)

REPORT . . .

1961 *of the Scheduled Areas and Scheduled Tribes Commission, 1960–61*, Delhi

REPORT . . .

1883 *on the Census of the Baroda Territories, 1881*, Bombay

REPORT . . .

1868 *on the Revenue Settlement Introduced into the Soopa Taluka*

REVISION . . .

1907 *Settlement of the Navsari Taluka of the Navsari Division, 1906–07*, Baroda

REVISION . . .

1911 *Settlement Report Palsana Taluka*

ROGERS, A.

1892 *The Land Revenue of Bombay*, London

ROWE, W. L.

1963 "Changing Rural Class Structure and the Jajmani System," in *H.O.*, vol. XXII, 41–44

ROYAL COMMISSION . . .

1927 on Agriculture in India, vol. II, pt. 1, *Evidence Taken in the Bombay Presidency*, London

SELECTION . . .

1887 *from the Letters, Despatches, and Other State Papers Preserved in the Bombay Secretariat*, Home Series, vol. I, G. W. Forrest (ed.), Bombay

SEN, B.

1962 *Evolution of Agrarian Relations in India*, New Delhi

SHAH, C. H.

1952 *Effects of World War II on Agriculture in India (with Special Reference to Gujarat)*, Ph.D. thesis, Bombay

SHAH, P. G.

1958 *The Dublas of Gujarat*, Delhi

SHAH, S. M.

1952 *Rural Class Structure in India, with Special Reference to Gujarat*, Ph.D. thesis, Bombay

SHIRRAS, G. F.

1924 *Report on an Enquiry into Agricultural Wages in the Bombay Presidency*, Government of Bombay Labour Office, Bombay

SHUKLA, J. B.

1937 *Life and Labour in a Gujarat Taluka*, Calcutta

SINGH, B.

1960 "Plan Approach to Agricultural Workers," in *Current Problems of Labour in India*, Delhi, 60–64

SINGH, M.

1947 *The Depressed Classes: Their Economic and Social Condition*, Bombay

SIVASWAMY, K. G.

1948 "Serf Labour among the Aboriginals," in *I.J.S.W.*, vol. VIII, no. 4

SIVERTSEN, D.

1963 *When Caste Barriers Fall: A Study of Social and Economic Change in a South Indian Village*, New York

SJOBERG, G.

1965 *The Preindustrial City: Past and Present* (paperback ed.), New York

SOLANKY, A. N.

1955 *The Dhodhias*, Ph.D. thesis, Bombay

SRINIVAS, M. N. AND A. M. SHAH

1960 "The Myth of Self-Sufficiency of the Indian Village," in *E.W.*, September 10, 375ff.

SRINIVAS, M. N.

1955 "The Social System of a Mysore Village," in McKim Marriott (ed.), *Village India*, Chicago, 1–35

1965 *Religion and Society among the Coorgs of South India*, reprint, London

1966 *Social Change in Modern India*, University of California Press

SYMINGTON, D.

Report on the Aboriginal and Hill Tribes of the Partially Excluded Areas in the Provinces of Bombay, Bombay

THORNER, D. AND A. THORNER

1962 *Land and Labour in India*, London

VEEN, K. W. VAN DER

1972 *I Give Thee My Daughter*, Assen

VREEDE-DE STUERS, S.C.L.

1965 "De Hindoe-maatschappij in overgang" [Hindu society in transition], in P. van Emst (ed.), *Panorama der Volken*, pt. 2, 263–358, Roermond

VYAS, V. S. (ED.)

1964 *Agricultural Labour in Four Indian Villages*, Sardar Patel University, Vallabhai Vidyanagar

WALLACE, R.

1863 *The Guicowar and His Relations with the British Government*, Bombay

WATSON, J. W.

1886 *History of Gujarat*, Bombay

WEBER, M.

1892 *Die Verhältnisse der Landarbeiter im ostelbischen Deutschland*, Leipzig

1922 "Wirtschaft und Gesellschaft," in *Grundriss der Sozialökonomik*, III Abteilung, Tübingen

1924 "Die ländliche Arbeitsverfassung," in *Gesammelte Aufsätze zur Sozial- und Wirtschaftsgeschichte*, Tübingen

1967 *The Religion of India: The Sociology of Hinduism and Buddhism*, transl. by H. H. Gerth and D. Martindale (paperback ed.), New York

WISER, W. H.

1936 *The Hindu Jajmani System*, Lucknow

WITTFOGEL, K.

1963 *Oriental Despotism: A Comparative Study of Total Power* (paperback ed.), Yale University Press

WOLF, E. R.

1966 *Peasants*, Englewood Cliffs, New Jersey

Index